P,

Cç

E

THE WONDERFUL WORLD OF
SIMULATION

a brief history of
modeling and simulation and
its impact on our lives

HENRY (HANK) OKRASKI
SES (RET), P.E., CMSP

Designed by Adina Cucicov at Flamingo Designs

The Library of Congress has cataloged the hardback edition as follows:

Okraski, Henry C.
 The Wonderful World of Simulation
 / Henry C. Okraski.—1st ed.
 1. Simulation 2. Training and Education 3. Science & Technology
 I. Henry C. Okraski II. Title

ISBN 978-1-938590-04-7

TABLE OF CONTENTS

FOREWORD

My parents, friends and family would frequently ask me what I did at work and I would explain that I was engaged in "simulation". Often their eyes would glaze over, not wanting to hurt my feelings by probing deeper. That was years ago when simulation was in its infancy and only understood by a rather small segment of society, but today we see examples in our everyday lives. Simulation might be termed the new calculus in analysis applications or it might be the key to education and training or it might be the added excitement in entertainment. Simulation is many things to many people. To me, it has been my total professional life.

This is the story about simulation—yet it is my story as I lived it. What I have attempted to do in this book is to answer the question put to me earlier by presenting the wide world of simulation as I know it. I purposely include many examples of simulators, some I have worked on and others that I am aware of.

I want to dedicate this writing to those brave folks who work in a field that is not widely understood in its totality—the men and women working in simulation applications involving the military, homeland security, education, medical and health care, transportation, entertainment, etc.

I also hope that, by reading this book, people can gain an appreciation for the career opportunities in a very exciting and stimulating field and an understanding of why mathematics and science (and art) are important in pre-paring for those careers.

Let me summarize what this book is all about. We begin this work as simu-lation entered my life and there is a thread of my career throughout the book. Of course, simulation has to be defined and people who have made a contribution to our field are recognized. Then, there is a journey through history with plenty of examples of historical simulators. There are many fig-ures in this book and that was intentional to give the reader an appreciation for the way technology has changed over the years and how imagination has been brought to bear on simulation applications. Several applications of simulators are presented and the expansion of the technology becomes obvious. I have only touched the surface of the *Wide World of Simulation*. Lastly, the future of simulation is described by two well-respected experts in their fields. I add my own projections.

I hope the reader gets a feel for the fulfillment I experienced throughout my career. When I reflect on the fact that the simulators that I have worked on have touched every serviceman and servicewoman, saving lives, time and money, it is very gratifying. My heartfelt thanks to all of those who I have worked with over the years who have made my work enjoyable and pro-ductive, particularly the men and women at the Naval Air Warfare Center Training Systems Division in Orlando, Florida where dreams become real-ity every day.

INTRODUCTION

My first experience with flight simulation was in Air Force flight training nearly 40 years ago. The training sessions were not very exciting and I must admit I did not look forward to time "under the hood". The technology was far from mature with low fidelity visual displays, and the simulator just did not feel or act like the real airplane. The F-4 Phantom II simulator, although state-of-the-art, was really just a part task trainer, a platform to practice emergency procedures and instrument approaches. But in the early 1980s, I was fortunate to get into the ground floor of the F-16 Fighting Falcon program as the Air Force retired the F-4. The transition to the F-16 was a giant leap into the digital age—advances in the power of computer processors had allowed industry to significantly improve the fidelity of the F-16 flight simulator and others. Even back then you could see what I would call an "evolution in the quality of simulation". And today the quality of simulation just continues to improve. For example, through distributed mission operations, simulators can be networked together to practice complex multi-aircraft formations and tactics. Simulators can be linked to live training exercises anywhere in the world and integrated into the exercise scenarios alongside real aircraft. This advance in fidelity and realism allows aircrew members to practice in a safe environment almost everything they must do in the real aircraft. In fact, due to range space limitations there are certain tasks pilots in the F-22 Raptor and F-35 Lightning II advanced fighters can only practice in the flight simulator. Simulation is no longer just about practicing routine tasks; simulators play a vital role in maintaining the readiness of America's warfighters. And simulation is no longer just focused on aviation training—virtually every specialty across the four military branches utilizes simulation as part of their initial and refresher training.

Hank Okraski's book chronicles the early evolution of simulation and then goes on to describe the *"Wonderful World of Simulation"* by exploring applications beyond defense that have emerged over the past several years. Simulation remains integral to defense, but is now applied to homeland security, entertainment, medical and health care, transportation...and the list continues to grow. With more than 50 years of simulation experience in government, industry and academia, Hank is uniquely qualified to offer the reader a comprehensive perspective. This is the only book I am aware of that covers simulation with the breadth and depth it richly deserves.

Simulation saves time, lives and money. "Modeling and Simulation" has been identified as a "critical technology" by the United States House of Representatives. And yet, I would guess most of America's population has no idea what simulation is all about. Our young people should be aware of this technology and understand what career opportunities are potentially available to them. And they should understand the importance of science and mathematics education as the building blocks for career opportunities in modeling and simulation. Hank hopes this book will stimulate the interests of our young people toward simulation careers, re-energize those of us currently working in the field and bring back cherished memories for those who paved the way in the early days of simulation.

Tom Baptiste, Lt Gen, USAF (Ret)
President/ Executive Director
National Center for Simulation

Chapter I

SIMULATION ENTERED MY LIFE EARLY

"Age is not a scalar quantity—life is a vector with magnitude and direction, with direction outweighing magnitude"

...the author

grew up in Utica, New York, attended local schools and went on to graduate from Clarkson University in Potsdam, New York with a degree in electrical engineering. When I was a little boy, I didn't live on the other side of the tracks: I lived <u>on</u> the tracks. The train tracks went right down the middle of Schuyler Street in a blue collar and no collar neighborhood where I lived, with the train having convenient access to the Utica Club Brewery across the street. As a consequence, there was plenty of noise throughout the night with railroad cars being switched, humped and banged around to add beer- laden cars to the trains. Engines were coal burners that "chugged" and spewed out loud steam with gusto. In Utica, the summers were hot and there was no air conditioning in any of the homes. In fact, we didn't even have a refrigerator. The ice man would deliver ice to us by carrying the 25 or 50 pound block on his shoulder destined for our kitchen "icebox". We slept with the windows open, the train engine and banging cars felt like they were in the apartment. When the train got moving, the vibration was felt throughout the rental flat. The fillings in your teeth became loose. Also, with the windows open, the soot from the train eked into the house, giving the furniture and your clothing a fine coat of that black stuff. Wonder we all didn't suffer from black lung disease. With the windows open, we all knew each other's business—more than we wanted. Also, the aroma of food cooking spread through the neighborhood. Mrs. Shabilski's kielbasa and cabbage cooking on her stove reminded us it was Saturday. She always had kielbasa on Saturday. Those were the best of times and the worst of times.

To get me out of the city in the summers, my parents would send me to my grandparents' dairy farm out in the country between Poland and Newport, New York. Part of their motivation was to get me out of the summer heat, but I really believe they wanted to stash me in a place where I couldn't get into much trouble. With both parents working, there were plenty of opportunities to get into mischief. I had plenty of company in that regard with the other kids in our neighborhood. They operated in "informal" gangs seeking out some questionable adventures. We didn't get into big trouble but some were border line activities. I was an only child and a poster "latch key" kid. My free time was spent climbing onto garage roofs and entertaining the racing pigeons kept in coops sitting on top of those garages.

There was no television in those days, so a lot of time was spent at the West Utica Boys Club and entertainment came from movies and the radio. We also collected, traded and played games with baseball cards. A favorite game was to flip your card at a wall and the player closest to the wall got to keep all of the cards in the game. We were avid comic book readers and collectors. I believe that's how we learned to read as well as we did. The comics sharpened our imaginations and provided experiences of adventure and fair play. The "good guys" always came out on top. My favorite was "Blackhawk". I still have a few of those comics and pick them up and read them occasionally. Radio was the source and center of our information and entertainment. At high noon, Kate Smith sang "God Bless America" every day for Jell-O—with the "big red letters". Radio had been dubbed as "the theater of the mind" and one can understand why because one could mentally visualize the happenings being broadcast. Radio really exercised the imagination and there were plenty of programs to rivet us youngsters to the Philco or Emerson radio. We had exciting programs such as Sky King, Sgt. Preston of the Northwest Mounted Police, Terry and the Pirates, Lone Ranger, The Shadow, Captain Midnight, Superman, Cisco Kid and Jack Armstrong—The All American Boy that had continuing adventures.

Most of these heroes' programs offered premium items. Such items as gun flashlights, hand-launched parachutes, pocket knives, glow-in-the-dark rings, and even a simulated Norden Bombsight were advertised. These were not cheaply made items. Not like today's products that are poorly made offshore. These were quality novelties that could take the punishment of a nine year old. With a box top from Wheaties or the label from Ovaltine, accompanied with ten cents, one could get a decoder ring or other item. The hero of the episode would give us a coded message on the air that we would have to decode. We listened attentively, writing down the letters as though we were FBI agents. The unfortunate child without a decoder ring was totally out of luck! Those of us with decoders were like members of a secret cult, not sharing the secret message with the unfortunates. Secrecy, spies and intrigue were very much "in" during and after World War II. These were the Golden Years of radio that provided entertainment and virtual experiences for millions of youngsters. When coupling imagination with an impressionable mind there are no limits to creative thinking.

As a premium, Jack Armstrong offered a set of simulated models that were prized by the kids. (See Figure 1).

Figure 1—Wheaties Models
(www.wheatieswings.com)

In 1944 during WWII The makers of Wheaties, General Mills Company, gave thousands of these models to our servicemen to build and fly while they recovered from their wartime injuries.

These WWII Wheaties model airplanes, the most popular fighters of the war, were also part of the Jack Armstrong Radio Show where the models were offered as Radio Premiums.

As these were constructed, the builder was required to put a penny in the nose for a nose weight. These models flew so well that National Championship Contests were organized. Awards were presented for best appearance, best maneuverability, and longest flight.

Wheaties had a "Secret" bomb sight which was supposed to emulate the Norden bomb sight which, by the way, was classified. (Figure 2)

Figure 2—Norden Bomb Sight Simulator (1942)
(www.worthpoint.com)

Immensely popular, the bombsight is a wooden replica of the famous Norton Bomb Sight. The body of the piece is covered by a paper label with 5 simulated gauges and dials--captioned, for example, "Altitude Adjuster... Numbers In Thousands Of Feet." On the narrow end is a green wooden revolving cover which allows the loading of three red wooden bombs. Another edge of the piece has a green wooden viewing tube; and when looked into, there is a pair of cross hairs over a mirror that reflects whatever is below the hole on the underside of the device.

The bomb launching is achieved by turning the green wooden revolving cover to drop each of the three bombs individually. An "Official Instruction Manual" has 15 different ship and submarine targets that could be cut-out.

It just so happens that my late father-in-law, Captain Ralph Dentinger, worked on the Norden Bombsight in World War II while on duty in England and received the bronze star for his improvements to the bombsight. I'll bet he didn't realize there was a Norden simulator available from Wheaties. Furthermore, I'd wager that the Germans didn't know that either!

One of the specialty products advertised on Jack Armstrong was a flight simulator. Wow, this really excited me because it represented an opportunity
to actually "fly "an airplane. I say actually "fly" because when the simulator
was linked to the narrative being broadcast on the radio, you felt like you
were in the air. Remember, this took place during World War II where pilots
were "white scarf" heroes and we enjoyed hearing about their accomplishments as in dog fighting enemy aircraft. Anyway, I sent in my box top from
Wheaties (which I was not terribly fond of) along with 25cents and greeted
the postman each and every day. The rural delivery postman would drop off
the mail and, shaking his head, said, "nothing for you today, Sonny". Finally,
perhaps a long two weeks later, the mailman handed the anxious nine year
old boy a large envelope. There was no doubt what was in the package-a real
simulator! The complete kit included the Pilot Trainer, Flight Manual, Corps
Wings, an Official Insignia and the Cub Pilot News. At that time, Piper Aircraft teamed with Wheaties to offer the kit that, no doubt, stoked the interest
of many youngsters into pursuing a pilot's license and maybe, a few years
later, purchasing or leasing a Piper Cub airplane. I had to assemble the device
which was not too difficult. Finally, I had a cardboard instrument panel, stick
and throttle and rudder pedals too. (Figures 3 and 4).

Figure 3—Artist redendering of the author at Nine Years Old
with His First Simulator (Dave Parker)

Figure 4—Jack Armstrong's Flight Simulator Kit
(www.worthpoint.com)

Just in time—Jack Armstrong's episode was about to be broadcast on the radio. Running into the house and turning on the radio, I sat erect in my chair, simulator in front of me and I was more than ready for my instructions and simulated flight. The radio character described where we would be flying. He said we were flying over a jungle. "Look down", he said, and I responded and, believe it or not, I could see the jungle! It was very green with a crystal river running through it. Imagination totally took over my brain as I "flew" my aircraft as per the instructions I was given by the radio character. To add to the realism, there was a tiny aircraft embedded in the simulator that moved as I moved the control stick. I could visualize the ground below in my mind's eye, even smell the rainforest. When the episode came to an end, I couldn't wait for the next in the series. I was "hooked" on simulation but didn't know at the time that I would spend more than 50 years working in simulation for the U.S.Navy and supporting the other military services.

Later in life, I was to learn that I was the victim of simulation "psychological" fidelity. I suspended disbelief long enough to be captured by the scenario that I was a part of. How wonderful! Today, children do not have the opportunities to exercise their imaginations like we did. Video games, videos, movies, television and theme parks pretty much hand the scenarios to the child with high fidelity graphics and sound. They are entertained without having to invest any cerebral energy in solving problems or "connecting the dots". How unfortunate that children do not have the valued experiences we had in the 1940's.

Anyway, that was my first experience with simulation. Many examples of simulation and specific simulators used for various purposes will be illustrated later in the book. I will, however, put more emphasis on simulators used for training purposes because that is what I know best. I will take you on a historical journey of simulators from the very beginning to the present time—bringing in, from time to time, my personal experiences.

Chapter II

WHAT IS SIMULATION?

A world of make believe and suspended disbelief…

There are various definitions of simulation. Some have said that anything that is not real is simulated. Under that definition, even "eggbeaters" are simulated eggs. Plastic daffodils are simulated flowers, etc.

A less broad definition might indicate that man-made artifacts including maps, statues, drawings, etc. are also simulations. Airplane models used by the military for recognition training are simulated aircraft. Even children's toys can be called simulators if you choose to call them that. One could simplify further by calling simulations "fakes". In the movie "Tora!Tora!", the Japanese pilots practiced their bombing attacks by sitting in the airplane while aircraft carrier crewmen dragged aerial maps of the Hawaiian Islands along the deck, under the airplane, with the pilot looking out of the side of the cockpit. That was how they did mission rehearsal!

Let us turn to Wikepedia for definitions as I often do in this book:

"Simulation is the imitation of the operation of a real-world process or system over time. The act of simulating something first requires that a model be developed; this model represents the key characteristics or behaviors of the selected physical or abstract system or process. The model represents the system itself, whereas the simulation represents the operation of the system over time. Simulation is used in many contexts, such as simulation of technology for performance optimization, safety engineering, testing, training, education, and video games. Training simulators include flight simulators for training aircraft pilots to provide them with a lifelike experience. Simulation is also used with scientific modelling of natural systems or human systems to gain insight into their functioning. Simulation can be used to show the eventual real effects of alternative conditions and courses of action. Simulation is also used when the real system cannot be engaged, because it may not be accessible, or it may be dangerous or unacceptable to engage, or it is being designed but not yet built, or it may simply not exist.

Key issues in simulation include acquisition of valid source information about the relevant selection of key characteristics and behaviours, the use of simplifying approximations and assumptions within the simulation, and fidelity and validity of the simulation outcomes." (*Wikepedia)*

Simulations begin with mathematical models and are put into a form whereby a computer can make what is being modeled behave just like the item or system being simulated. Thus, we have computer-based simulations. And that is what we will be addressing in this book-primarily simulations that are physics-based and are used for training, analysis, education, entertainment and test and evaluation.

People are aware of flight simulators and have some idea how they are used by the military, airlines and NASA for training. However, simulation technology is really shaping our lives and we may not be aware of its many applications. Hardly a week goes by where an accident or other situation is animated on television to give the viewers some better description, using visualization. When the brave airline pilot, "Sully" Sullenberger, landed on the Hudson River, it was possible to play back the events to and including setting the plane down on the water without loss of life. It was also possible to hear the airplane crew and control tower exchanging information just as they were during the actual flight mishap. Simulation technology has allowed us to better understand our world in so many areas. Commercial products such as computer games have taken simulation from the interest of a small community to the general population. Laser Tag is a good example of a military technology that has been "spun-off" to the commercial gaming market. I was privileged to have worked with the inventor of the Multiple Integrated Laser Engagement System (MILES), Al Marshall was one of the most creative engineers in our business of all time. (Figure 5). He brought MILES to the Army and Marine Corps. (Figure 6). Al was a brilliant inventor who had no stomach for politics or red tape, and he rarely took "no" for an answer. He was a remarkable individual. He was only one of the Navy's many "imagineers" working to develop more effective tools for training. Fortunately, he had a great team supporting him.

Figure 5—MILES Inventor Al Marshall and Author (left to right)

Figure 6—An Urban Exercise using MILES Technology
(note Laser Detectors)—Wikipedia Photo

Back to the definitions: the military breaks down simulation further into three areas: live, virtual and constructive (LVC).

The LVC categories are defined as follows: *(Wikipedia-Live, Virtual, Constructive)*

• **"Live**—Modeling and Simulation (M&S) involving real people operating real systems, e.g. a pilot flying a jet.

• **Virtual**—M&S involving real people operating simulated systems. Virtual simulations inject a Human-in-the-Loop into a central role by exercising motor control skills (e.g., flying a simulated jet), decision making skills (e.g., committing fire control resources to action), or communication skills (e.g., as members of a C4I team).

• **Constructive**—M&S involving simulated people operating simulated systems. Real people stimulate (provide inputs) to such simulations, but are not involved in determining the outcomes.

Our definitions would not be complete if we did not include simulation interoperability.

- **Interoperability:** the ability to exchange usable data between two or more systems and to invoke their services using the appropriate input parameters!"

Interoperability is the feature of simulation that enables, for example, a simulator in Norfolk, Virginia to be connected with another simulator, perhaps of a different type at Cherry Point, North Carolina, enabling the conduct of a training exercise in the same battle space. The ability to conduct team training in simulated exercises has opened the door of opportunity to "train as we fight and fight as we train."

Chapter III

THE PEOPLE WHO MADE A DIFFERENCE IN SIMULATION

Simulation is not a profession-it is a way of life
...Janos Sebestyen Janosy

I t is very difficult to identify all of the individuals who have made what might be called significant contributions over time to simulation technology. Perhaps fools rush in where wise men fear to tread but I will put my cards on the table by listing those explorers and pioneers of our business. (Figure 7)

*Figure 7—Individuals who have made significant contributions to M&S
(H.Okraski)*

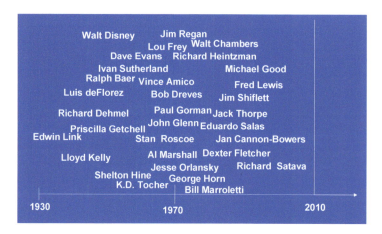

I am sure there are more who should be added to this first pass at a list. Again, most of my experience is in real-time virtual simulators. So, you might discover that my list is somewhat biased. This figure attempts to identify those individuals that made significant contributions to modeling and simulation as we know it today.

There are several military leaders who I personally observed that strongly advocated the use and set in motion policies to institutionalize simulation. I dare not attempt to list all of them here for fear of inadvertently omitting some of them. A few of the most obvious are RADM Luis de Florez, VADM Malcolm Cagle, VADM John Disher, VADM Al Harms, RADM Fred Lewis, Gen Paul Gorman, LtGEN John Jumper, BGEN Jim Ball and BGEN Steve Seay.

Numerous civilians contributed to the successful growth of simulation. Some of them are included in this writing. Again, I hesitate on listing these people here.

There are three individuals who are the backbone of modern modeling and simulation. Without these pioneers in our field we would be somewhere in the dark ages of simulation.

Luis de Florez, 1889-1962

"Luis de Florez was born in New York City. De Florez attended MIT, and graduated in 1911 with a B.S. in Mechanical Engineering. De Florez worked in the United States Navy as a career officer in World War I. He worked in the aviation section of the Navy and also on the development of refinery technology.

In the 1930s, de Florez also worked as an engineering consultant for various oil companies. His name is on several patents, including a 1918 U.S. patent (#1,264,374) for a "Liquid prism device with rigid closed sides" which included a system for varying the density of a medium filling the prism and thereby varying the refraction of light waves passing through the prism, and a 1930 Canadian patent for the "cracking and distillation of hydrocarbon oils". During World War II, he gave up his business to help solve the Navy's training problems.

In 1941, then Commander de Florez visited the United Kingdom and wrote what would become an influential report on British aircraft simulator techniques. It influenced the establishing of the Special Devices Division of the Navy's Bureau of Aeronautics (what would later become the Naval Air Warfare Center Training Systems Division (NAWCTSD). Later that year, Commander de Florez became head of the new Special Devices Desk in the Engineering Division of the Navy's Bureau of Aeronautics. De Florez championed the use of "synthetic training devices" and urged the Navy to undertake development of such devices to increase readiness. He also worked on the development of antisubmarine devices. De Florez has been credted with over sixty inventions.

During World War II, he was subsequently promoted to captain and then to Flag rank, becoming a rear admiral in 1944. In 1944, de Florez was awarded the Robert J. Collier Trophy for 1943 for his work in training combat pilots and flight crews through the development of inexpensive synthetic devices. De Florez was awarded with the Legion of Merit in June 1945. In 1946, Tufts University awarded de Florez an honorary Doctor of Science degree at commencement.

Admiral de Florez was the first director of technical research at the CIA. In the mid-1950s, de Florez was the president of the Flight Safety Foundation. Presented since 1966, the Foundation's *Admiral Luis de Florez Flight Safety Award* is named after him. It recognizes "outstanding individual contributions to aviation safety, through basic design, device or practice." De Florez established a trust to support the award that provides each recipient with $1,000. Luis de Florez died in November 1962, at the age of 73 in the cockpit of his airplane, which was ready for take-off at a Connecticut airport. The main building complex at the **Naval Air Warfare Center Training Systems Division, Naval Support Activity Orlando, Florida, is named in his honor."** (*Wikipedia*)

A little known fact is that in 1939, Luis de Florez was given an honorary doctorate by Rollins College of Winter Park, Florida. (Figure 8)

Figure 8—Luis de Florez receiving Honorary Doctorate Degree
at Rollins College 1939 (Second from right)
(Marjorie Kinnan Rawlings is fifth from left)

(Marjorie Kinnan Rawlings Papers, Department of Special and Area
Studies Collections, George A. Smathers Libraries, University of Florida)

In 1939, de Florez was awarded an honorary doctorate degree from Rollins College in Winter Park, Florida. His area of expertise at that time was not simulation. It had to do with his work in the oil industry and other research areas. Marjorie Kennin Rawlings, author of "The Yearling" and other works, received a doctorate at the same time. (She is pictured fifth from the left in the figure above.)

Edwin A. Link, 1904-1981

""Ed" Link was born in Huntington, Indiana, but moved in 1910 to Binghamton, New York, where his father purchased a bankrupt music firm. The business was renamed the Link Player Piano and Organ Company and it was here where Ed would begin and develop his multi-faceted career as an inventor, industrialist and pioneer in the fields of flight simulation, underwater archaeology and ocean engineering. To quote his friend

Harvey Roehl, Edwin A. Link, Jr. was a "backyard inventor in the finest American sense."

In his early twenties, at considerable expense and some risk, Edwin A. Link, Jr. obtained his pilot's license. While struggling to become a pilot, he began tinkering with parts of organs at his father's factory, trying to develop a training device so that pilots could start learning to fly safely and inexpensively without leaving the ground. Initially his Link Trainer, although successful, was seen as a toy and relegated to the status of fairground ride.

In the mid-1930's, after a series of air accidents, the Army Air Corps ordered six of Link's instrument trainers to enhance its pilot training program. Once public attention had been drawn to this practical device, orders for more came from all over the world. Ultimately Link's flight trainer (the Blue Box) led to the development of the whole field of flight simulation. With the help of his wife, Marion Clayton Link, whom he had married in 1931, Ed Link ran a highly successful enterprise, Link Aviation, Inc., throughout World War II and until he sold the company in 1954.

Thereafter Ed's skills and attention focused on underwater archaeology and exploration. In this, his wife, Marion, became his partner in research, and, with their two sons, William Martin and Edwin Clayton, they undertook a number of voyages. During these years, Ed Link worked constantly to improve diving equipment in order to allow divers to go deeper, stay longer underwater, explore more safely and efficiently, and return to the surface with less risk. On one of the sea voyages in 1973, during a routine dive in a submersible, the Links' younger son, Clayton, and his friend Albert Stover were killed. In a very moving statement to the press, Ed expressed his conviction that their death had not been in vain, but had identified problems which must be solved in order to meet the challenge of safer underwater exploration.

Link continued actively exploring, tinkering, writing and generally en-
joying his many interests until very shortly before his death in 1981. His
was an unusually generous spirit: not only did he give tirelessly of his
time and energy; he also donated financially to many foundations, scholarships and charitable causes. His gifts to Binghamton University include
the Edwin A. Link Organ Music Professorship, Binghamton's first en-
dowed faculty chair." *(Wikipedia)*

Richard Dehmel, Ph.D. 1904-1992

"Richard Dehmel, inventor of the Dehmel Flight Trainer/Simulator, was
granted U.S. patent No. 2,494,508 for Means for Aircraft Flight Training
on Jan. 10, 1950. The invention was the first to solve the equations of
flight and have the controls and instruments of the trainer respond as an
accurate equivalent of a real airplane.

The trainer/simulator dramatically reduced the cost, time and risk to
train aircraft crews. It also allowed a significantly higher level of training
in „extraordinary situations. For example, Pan American World Airways
trained 125 flight crews, plus 46 British Overseas Airways and 85 military
transport crews during 13,000 hours of simulator time. The simulator
enabled Pan Am to reduce crew training costs by 60 percent and in-flight
training time from 21 to eight hours per crew.

Dehmel spent much time building a multi-talented team to accelerate the
design and production of this critical tool for use during the latter half of
World War II.

Dehmel earned masters and doctorate degrees from Columbia University
after earning a mechanical engineering degree from the University of
California. He was a 1991 inductee into the Aviation Hall of Fame. *(New
Jersey Inventors Hall of Fame)*

- Chapter IV -

EARLY YEARS OF SIMULATION

From abstraction to near-reality..

t is difficult to say with certainty when simulation was born. Of course, it depends on how simulation is defined. In the broader sense, simulation could include the entire universe of abstraction. This would include statues, maps, drawings, etc. Looking way back, one could imagine carvings on a cave wall occupied by Og, the Neanderthal. These primitive drawings could reflect the experiences of the cave dweller, depicting animals, fish, etc.. These scrawlings are a form of simulation. It might also be theorized that Og may have constructed a "simulator" to keep his neighbor cave dweller, Da, from raiding his fresh kills. I can visualize Og setting up a pile of stones, much like a snowman, and swinging his trusty club at the top stone, pretending it was that nasty Da's head. A little practice on that rock simulator and Og was confident that he could make that first swing effective. Tongue in cheek, simulation played a role in evolution.

Imaginations led to all sorts of simulator concepts. For example, as civilization progressed and it was necessary to band together to take on bigger or more game, Og and his friends felt they had to perform as a team and no longer could hunt individually. So, the need for some form of team trainer became evident. Below is a cartoon representing an early attempt at team training using a "low fidelity" trainer. (Figure 9).

Figure 9—Prehistoric Men Engaged in Early Team Training
(Dave Parker)

Carved Maps

Getting back to more documented examples of abstraction, maps have been used for thousands of years. We take maps for granted today and rely heavily on GPS for navigation. The Babylonians had maps carved into stones as far back as 3,000BC. Imagine the difficulty in that stage of human development in translating the real world to a stone carving. Fundamentally, how does stone equal information? (Figure 10).

Figure 10—Babylonian Map in Stone (Wikipedia)

Stick Charts

Another interesting example of an abstract device that served the Polynesians as a training device and was also used as a navigation aid is the Mattang. (Figure11)

These "stick charts", as they were called, took on various shapes and sizes.

Figure 11—Mattang Navigation Trainer (Wikipedia)

Stick charts were made and used by the Marshallese to navigate the Pacific Ocean by canoe off the coast of the Marshall Islands. The charts represented major ocean swell patterns and the ways the islands disrupted those patterns, typically determined by sensing disruptions in ocean swells by islands during sea navigation. Stick charts were typically made from the midribs of coconut fronds tied together to form an open framework. Island locations were represented by shells tied to the framework, or by the lashed junction of two or more sticks. The threads represented prevailing ocean surface wave-crests and directions they took as they approached islands and met other similar wave-crests formed by the ebb and flow of breakers. Individual charts varied so much in form and interpretation that the individual navigator who made the chart was the only person who could fully interpret and use it. Use of stick charts and navigation by swells apparently ended after World War II, when new electronic technologies made navigation more accessible, and travel between islands by canoe lessened. *(Wikipedia—Marshall Islands Stick Charts).*

Puppets

Puppets originated in India at about 1,000BC where they were used to act out serious epics like Maha Bharata and Bala Ramayana. These were considered to be sacred productions although there was an element of entertainment to them. Later, the Japanese adopted human-like puppets to replace human actors. It is said that the "producers" of the shows got tired of paying actors and decided to use puppets-who were paid a lot less! Of course, in modern times, who is not familiar with Jim Hensen's Muppets. Today, puppets are finding their way helping kids cope with such situations as "bullying", child abuse and other topics that the children are hesitant to talk about but, through puppets, they find it much easier to communicate.

Given this quick snapshot of abstracts, let us now move forward to the time when knights were bold. Practice meant the difference between life and death-or winning the hand of the fair maiden or losing one's head. The Quintain Trainer provided instant feedback to the trainee! (Figure 12)

Quintain Trainer

Figure 12—Quintain Trainer (Dick Adkins)

The Middle Ages was an extremely violent era in history featuring battles in both Europe and the Holy Land when the crusades, and the crusaders who fought them, were numerous. Feudal Lords and Knights were expected to be expert in the use of medieval weapons—pell training was essential. The quest for power led to invasions of lands and territories which had to be fought for. Warfare during the Middle Ages, or Medieval era called for a variety of weapon expertise. Knights and men-at-arms, or foot soldiers, used a variety of different weapons and they had to be skilled in the use of all of them. Pell Training (Quintain) was predominantly used by a Knight but other soldiers would also practice at the pell. Knights also had to have additional weapon training—use of the lance was practiced extensively. *(www. middle-ages.org)*

Moving to more modern examples, there were several attempts at constructing flight training devices. The interest in aviation stems from the novel concept of man actually being able to fly. It was a new and seemingly boundless frontier. But to fly meant that a person had to learn the procedures and be prepared for any situation that might occur. Failure to do so could cost the pilot his or her life. Some examples of the innovations created through necessity are as follows:

Gunnery Trainer

They say that history is merely the biographies of great people. One such individual was Juri Vladimirovich Gilsher, a Russian flying ace who was in a terrible crash of a Sikorsky S-16 where he lost his left leg. Brave as he was, he continued to fly and decided that there had to be a better way to teach gunnery other than in the actual aircraft. His invention taught new pilots how to fire the wing-mounted machine guns. (Figure 13) Gilsher's unit was fitted- out with the Nieuport 17, one of the most successful aircraft fighters in World War I. The first Nieuports had a Lewis machine gun mounted on the center of the upper wing, which the pilot could pull downwards and push upwards with the help of a support in order to be able to attack an enemy from below or above in the blind angle. These wing-mounted machine guns were, however, extremely hard to handle. As they say, necessity breeds invention as was the case with Juri Vladimirovich Gilsher, a Russian hero who continued to serve his country with flying skills and creativity. (*Walter Ulrich*)

Figure 13—Gilsher's gunnery trainer
(Kobelkov, from the Collection of Walter Ulrich)

Cockpit Crew Trainer

The reconnaissance mission of the German air arm consisted of the pilot and his observer who was positioned in the rear seat of the open cockpit aircraft. The observer's mission was to take photographs and keep notes on what he observed. Even in the early 1900's there was an issue of aircrew co-ordination. For example, when the pilot of the airplane made violent moves, unfortunately, the back seat observer would, at times, exit abruptly from the airplane and probably took the camera with him. (I don't know why they did not use seat belts at that time- but they did not.) A young first lieutenant named Carl Fink had a better idea. He set about to develop a simulator that may have been one of the first physiology trainers. He set up his trainer (Figure 14) at Doberitz, which was to the west of Berlin, and trained pilots and observers to coordinate aircraft movements and prepare for the forces that might be felt by the observer. The simulator included an engine and fuselage of the aircraft, sans wings, with the fuselage hung on a 3 meter pivot-mounted ring. The fuselage could be moved in pitch and yaw and the entire mass could be rotated a full 360 degrees. *(Walter Ulrich)*

Figure 14—Carl Fink's Cockpit Crew Trainer

(Credit: Biermann / Erhard Cielewicz From Collection of Walter Ulrich)

Flight Simulator with Hydraulics

The Germans also had a rather sophisticated apparatus constructed by the Berlin engineer and pilot Franz Drexler. (Figure 15). Back in 1913, this inventor had released a prototype for mechanising and automating flight stabilization of large aircraft. These early giants, long-range reconnaissance aircraft and bombers, with a wingspan exceeding 40 meters but with a relatively weak structure were difficult to maneuver. In 1916, Drexler produced the "Training Swing", a device which could be considered the first German flight simulator. Its purpose was to impart the "sensation of hydraulic steering movements" to the potential pilots of large aircraft. He brought in the man-in-the-loop in an otherwise automated process where an automat coupled, via a hydraulic system, the plane's gyroscope and the control surface. At the front of his device was an electric motor which replaced the impeller of the aircraft generator. Behind that was the control module, from which four control wires moved the swing relative to the base plate. The student pilot thus learned to steer the bulky airplane with the tip of his finger much like we steer our automobiles with power steering. It may have been what we call a "Rube Goldberg" invention today but it was effective as a training device. *(Walter Ulrich)*

Figure 15—Drexler's Training Swing

(Credit Deutsches Museum. From the Collection of Walter Ulrich)

Antoinette Barrel Trainer

The French had a novel trainer to familiarize pilots, or want-to-be pilots, with the sensations of aircraft movements. It was probably "the" technology of the time.

A training rig was developed in 1909 to help the pilot operate the control wheels before the aircraft was flown. (Figure 16). This consisted of a seat mounted in a half-barrel and the two wheels. The whole unit was pivoted so that assistants outside could pitch and roll the device in accordance with the pilot's use of the control, using long wooden rods attached to the barrel structure. A full-size model of the "Antoinette Barrel Trainer" is in the foyer of the Airbus Training Centre at Toulouse, France. *(Wikipedia Flight Simulator)*

Figure 16—Antoinette Simulator (Wikipedia)

Sanders Teacher Trainer

The Sanders Teacher Trainer was as close to flying an actual airplane as one might hope in 1910. The trainer was marketed with a focus on novices and those interested in experiencing actual flight. The Wright brothers made their historic flight in 1908 so the nation was energized on "flight." The Sanders Teacher Trainer was a modified airplane that was mounted on a universal joint (rocker) settled on the ground. The pilot could control the aircraft surface with his control stick and search for equilibrium in flight. It was necessary to have a good supply of wind to make it work, however. (Figure 17)

Figure 17—The Sanders Teacher Trainer

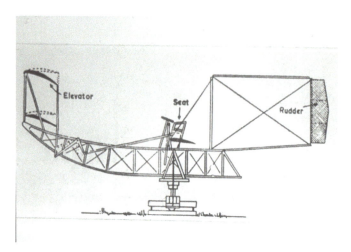

Ruggles Orientator

Today, there are a number of physiology trainers used to acquaint pilots and others with the forces on their bodies that they might experience in flight. There were early attempts to provide those experiences. The Ruggles Orientator, for example, consisted of a seat mounted within a gimbal ring assembly which enabled full rotation of the student in all three axes and, in addition, provided vertical movement. All motions were produced by electric motors controllable by the simulated sticks and rudder bars of the student and examiner. This device was stated to be useful for "developing and training the functions of the semi-circular canals and, incidentally, to provide such a machine for training aviators to accustom themselves to any possible position in which they may be moved by the action of an aeroplane while in flight. This device was also used for screening potential pilots. Quite simply, some people are more prone to motion sickness than others. It is better to determine this early in a pilot's career rather than later. A further optimistic claim was that the aviator could be blindfolded so that the sense of direction may be sensitized without the assistance of the visual senses. In this way, the aviator, when in fog or intense darkness, may be instinctively conscious of his position. A word of advice: do not enjoy a bowl of chili before subjecting yourself to this trainer. *(www.simulationinformation.com)*

Link "Blue Box"

The baseline for modern simulators used for training began with the Link "Blue Box". Over time, the somewhat primitive design yielded to user needs and applications. In our business, the mention of the "Blue Box" is usually greeted with a big smile and a measure of respect for the inventor. (Figure 18)

It all began in an organ factory. Can you imagine two more unconnected areas-aviation and organs? Today, we talk about thinking "outside of the box" but here we have an inventor who not only thought outside the box but invented a new one!

Figure 18—Original Blue Box on display at Western Canada Aviation Museum (Wikipedia Commons)

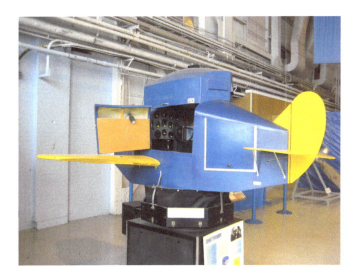

As discussed earlier, Edwin Link gained his early engineering experience with his father's firm, The Link Piano and Organ Company, of Binghamton, New York. His first patent was granted for an improvement in the mechanism of player pianos! The flight trainer was developed in the period of 1927-29 in the basement of the Link factory and he made use of the familiar vacuum/pneumatic mechanisms from the family business. The first trainer, advertised as 'An effective Aeronautical Training Aid—a novel, profitable amusement device'—was described in a patent filed in 1930. Pitch, roll and yaw movements were initiated in the same manner as in its predecessors, but vacuum/pneumatic bellows were used for actuation. Vacuum was also used as an analog computing medium for the instruments.

The device was not regarded as a wise investment by others at that point, nor seen to meet a training need.(In fact, his first trainers were found in amusement arcades as an entertainment thrill ride.) By now, a serious competitor was also on the scene in the simulator built by H.A. Roeder. A need certainly did exist in instrument training, however, so Edwin Link fitted cockpit instruments as standard equipment to his design and blind flying training was started by the Links at their flying school in the early 1930s. The importance of this training was soon recognized, notably by the U.S. Army Air Corps, when they took on responsibility for Air Mail delivery. This gave Link Trainers a great reputation with increasing sales.

The Model C followed in 1936, able to rotate through 360 degrees, which allowed for a magnetic compass to be installed, while the various instruments were operated either mechanically or by vacuum. With further refinements along the way, the Link Trainer became a simple form of analog computer, fitted with a full set of instruments to guide the pupil on an imaginary flight. The simulated course is automatically recorded and traced by the three-wheeled course plotter (the self-propelled and steerable "crab") across paper, or a map on the instructor's desk. A duplicated instrument panel is also present, electronically harmonized with those in the trainer's cockpit. This miniature airplane is pivoted on a universal joint mounted on an octagonal turntable, which in turn is free to rotate in azimuth on a square base. Between the fuselage and the turntable are four supporting bellows, which are inflated or deflated by a vacuum turbine. Its valves are operated as the pupil moves the control column, and realistically recreates most of the sensations and "feel" of flying.

Both calm and rough-air flying conditions could be created by the instructor. The trainer could also initiate stall when recorded airspeed and attitude fall outside pre-determined limits. It went into a very realistic spin, with the instruments performing normally for such conditions. A cross-country "flight" of up to 200 miles was possible, during which the instructor was able to confront the pupil with most of the difficulties that might occur during a genuine flight. Link Trainers were also sold to Russia, Japan, France and Germany—a Luftwaffe bomber pilot of 1940 had spent 50 hours in a Link Trainer!

Several models of Link Trainers were sold in a period ranging from 1934 through to the late 1950s. These trainers kept pace with the increased instrumentation and flight dynamics of aircraft of their period, but retained the electrical and pneumatic design fundamentals pioneered in the first Link.

Trainers built from 1934 up to the early 1940s had the a color scheme that featured a bright blue fuselage and yellow wings and tail sections. These wings and tail sections had control surfaces that actually moved in response to the pilot's movement of the rudder and stick. However, many trainers built during mid to late World War II did not have these wings and tail sections due to material shortages and critical manufacturing times.

The Link Trainer type D4 of 1950 is also designed and equipped to provide thorough instruction in the use of radio signals, radio beams and radio compass. Both the Lorenz Landing System or SBA and SCS51 precursor of ILS for blind landings in fog were simulated. Modern-day computerized flight simulators are much more sophisticated and very realistic—and also expensive. In its time, however, the Link Trainer was at the forefront of flight technology, saving many lives... and actual airplanes. *(Norfolk and Suffolk Aviation Museum)*

A short personal experience : A few years ago, I was asked to speak to the "Friends of the Library" at the Winter Park, Florida library. I was not overly excited about the prospect of doing so. I thought my talk would be perhaps too esoteric for the audience-mostly made up of elderly retirees, and we both might be bored. I began the presentation by describing the Link Trainer. When I got into the third sentence, a voice from the rear of the room exclaimed, "I used to work on that trainer in Binghamton! ". He was joined by another outburst from the opposite side of the room. "I designed the position locator for the Links", the elderly man said. I was blown away. The two gentlemen, now living in a retirement community in Winter Park had actually worked with Ed Link in the design of the trainer. So, what began in my mind as a "ho-hum" presentation turned out to be a memorable and fascinating evening.

ANT-18 Trainer

The ANT-18, a modified blue box, consists of two main components.

The first major component is the trainer itself. The trainer consists of a wooden box approximating the shape of a cockpit and forward fuselage section, which is connected via a universal joint to a base. Inside the cockpit is a single pilot's seat, primary and secondary aircraft controls, and a full suite of flight instruments. The base contains several complicated sets of air-driven bellows to simulate movement, a vacuum pump which both drives the bellows and provides input to a number of aircraft instruments, and a device known as a Telegon Oscillator, which controls the remaining instruments.

The second major component is an external instructor's station, which consists of a large map table, a repeated display of the main flight instruments, and a moving marker known as a "crab." The crab moves across the glass surface of the map table, plotting the pilot's track. The pilot and instructor can communicate with each other via headphones and microphones.

The ANT-18 has three main sets of bellows. One set of four bellows (one under each corner of the cockpit) controls movement in the pitch and roll planes. A very complicated set of bellows at the front of the cockpit controls movement in the yaw plane. This complex set of 10 bellows, two crank shafts and various gears and pulleys comprised the turning motor. This motor could turn the entire cockpit in continuous 360 degree circles. This was possible since a series of electrical slip ring contacts in the lower base compartment, supplied electrical continuity between the cockpit and the base.

A third set simulates vibration such as stall buffet. Both the trainer and the instructor's station are powered from standard 110VAC/240VAC power outlets via a transformer, with the bulk of internal wiring being low voltage. Simulator logic is all analog and is based around vacuum tubes.

The Link Trainer touched the lives of over one half million pilots in World War II alone. There are individuals who were close to Ed Link. One of them was the famous aviatrix, Amelia Earhart.

Amelia Earhart

Some time before Amelia Earhart attempted her unsuccessful and fatal flight around the world she became acquainted with another aviation enthusiast, Ed Link. Obviously, she was interested in instrument flying and navigation and she received some of that training in the Link Trainer. (Figures 19 and 20)

Figure 19—Amelia Earhart and Ed Link (Kelly-Pilot Maker)

Figure 20—Amelia Earhart in a Link Trainer (Kelly-Pilot Maker)

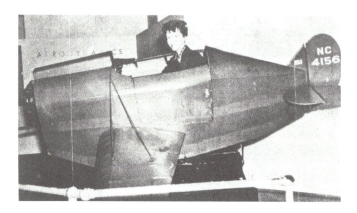

The disappearance of Amelia Earhart still remains a mystery but there are initiatives underway today to solve the case. For those who may not be familiar with her attempt to fly around the world and her disappearance, the following account of the search efforts is both informative and very interesting.

Beginning approximately one hour after Earhart's last recorded message, the United States Coast Guard Cruiser (USCGC) *Itasca* undertook an ultimately unsuccessful search north and west of Howland Island based on initial assumptions about transmissions from the aircraft. The United States Navy soon joined the search and over a period of about three days sent available resources to the search area in the vicinity of Howland Island. The initial search by the *Itasca* involved running up the 157/337 line of position to the NNW from Howland Island. The *Itasca* then searched the area to the immediate NE of the island, corresponding to the area, yet wider than the area searched to the NW. Based on bearings of several supposed Earhart radio transmissions, some of the search efforts were directed to a specific position on a line of 281 degrees (approximately northwest) from Howland Island without evidence of the flyers. Four days after Earhart's last verified radio transmission, on July 6, 1937, the captain of the battleship *Colorado* received orders from the Commandant, Fourteenth Naval District to take over all naval and coast guard units to coordinate search efforts.

Later search efforts were directed to the Phoenix Islands south of Howland Island. A week after the disappearance, naval aircraft from the Colorado flew over several islands in the group including Gardner Island (aka Nikumaroro Island), which had been uninhabited for over 40 years. The subsequent report on Gardner read: "Here signs of recent habitation were clearly visible but repeated circling and zooming failed to elicit any answering wave from possible inhabitants and it was finally taken for granted that none were there..." At the western end of the island a tramp steamer (of about 4000 tons)... lay high and almost dry head onto the coral beach with her back broken in two places. The lagoon at Gardner looked sufficiently deep and certainly large enough so that a seaplane or even an airboat could have landed or takenoff in any direction with little if any difficulty. Given a chance, it is believed that Miss Earhart could have landed her aircraft in this lagoon and swum or waded ashore. They also found that Gardner's shape and size as recorded on charts were wholly inaccurate. Other Navy search efforts were again directed north, west and southwest of Howland Island, based on a possibility the Electra had ditched in the ocean, was afloat, or that the aviators were in an emergency raft.

The official search efforts lasted until July 19, 1937. At $4 million, the air and sea search by the Navy and Coast Guard was the most costly and intensive in U.S. history up to that time, but search and rescue techniques during the era were rudimentary and some of the search was based on erroneous assumptions and flawed information. Official reporting of the search effort was influenced by individuals wary about how their roles in looking for an American hero might be reported by the press. Despite an unprecedented search by the United States Navy and Coast Guard, no physical evidence of Earhart, Noonan or the Electra 10E was found. The aircraft carrier USS *Lexington*, the *Colorado*, and the *Itasca* (and even two Japanese ships, the oceanographic survey vessel *Koshu* and auxiliary seaplane tender *Kamoi*) searched for six–seven days each, covering 150,000 square miles (390,000 km^2). Of note, one of the pilots flying off the Lexington in the search was Adm., John Madison Hoskins, a colorful World War II hero, who returned Charles Lindbergh via naval ship from Paris to New York after his crossing of the Atlantic from New York to Paris. Obviously, the aviators of the time, perhaps few and far between, formed a close-knit group. (*Wikipedia*)

- Chapter V -

WORLD WAR II AND
THE LINK TRAINER

The Greatest Generation earns their stripes...

When Ed Link first developed his trainer in his father's organ factory, the "Blue Box," he probably had no idea that it would play a major role as a weapon against the Axis Alliance. In fact, it is noted that a senior Canadian officer stated that the war against Germany was not won in the sky over Germany but rather in the Link trainer. Royal Canadian Air Force Chief of Staff, Air Marshall Robert Leckie, stated that "The Luftwaffe met its Waterloo on all the training fields of the free world where there was a battery of Link Trainers." *(Kelly).*

We know that Link Aviation of Binghamton, N.Y. produced about 10,000 Blue Boxes and, with them, we trained over 500,000 pilots. It is hard to imagine an air force that large. That would be about eight times the population of Utica, New York! The total number of commercial pilots today, according to USA Today, is about 90,000. Our country mobilized to develop the pilot pipeline using the Link Trainer as a production tool. The United States was not the only country to use the trainer, however. Trainers were also purchased before the war by Japan, Britain, France, Germany, Russia and others.

Given the presence of the trainer in so many places, I want to tell the stories of two U.S. Military units that were heavily involved with the Link Trainer. I also describe a third group of foreign aviators that may or may not have employed the Link Trainer in their training but all indications are that they did. In any event, I feel that their story must be told. It involves patriotic women aviators who bravely flew combat missions before the United States entered World War II.

Link Instructor/ Operators

The Navy Department was in serious trouble with having to provide sufficient instructor/operators for the thousands of Link Trainers that were fielded. Men were needed to fight the enemy and could not be diverted from their primary job. They turned to the WAVES (Women Accepted for Volunteer Emergency Service) for their support. The WAVES began a full-scale recruitment program to bring young women into the service to support flight training and other programs. Recruiters went to colleges and universities seeking female students majoring in education or communications who were willing to contribute to the war effort. In December 1942, the Navy's first female boot camp was formed on the Iowa State Teachers College campus in Cedar Falls. Approximately 84,000 WAVES were recruited over the next three years. Some would be assigned to pilot training-training pilots, that is. Twenty WAVES were sent to Atlanta to learn all about Link Trainers, the flight simulators in which Navy flight students learned instrument flying.

I had the pleasure of meeting one of those exceptional women. Priscilla (Pat) Getchell lives in Central Florida. She is a retired high school counselor and still does a considerable amount of volunteer work with Angel Flight Southeast. Angel Flight is a charitable organization comprised of pilots and others involved in aviation. Pat Getchell arranges flights for people in need who cannot afford to fly or are unable to do so. For example, if a child in Orlando needs to get to the burn center hospital in Houston, Texas, Pat seeks out volunteer aviators who will make that flight with the sick child. Today, she does the scheduling of these "mercy flights."

She was 86 years old when we had lunch. You would never have guessed her age. She is an exciting lady and, more importantly, excited about life. A few years ago the National Center for Simulation (NCS) honored her at a general meeting for her service to the country. It was a celebration long delayed and much deserved.

During World War II, the nation was escalating its air forces at an astounding pace. The WAVE team would step in to provide the Link portion of ground school training. The initial cadre of volunteers was to report to Atlanta, Georgia for Link Training. But when its members arrived at their Atlanta base, there were no accommodations for women, Pat Getchell reports. Consequently, they were billeted on the fifth floor of the Biltmore Hotel in downtown Atlanta. Not a bad assignment, she thought.

Some definition of what the Link operator/instructor does might be appropriate at this time: the Link Operator sits at a desk and, using a radio, gives signals to the pilot sitting in the Link Trainer, enabling him to navigate from one radio station to another and then to a Control Tower for landing. (Figures 21-28) They studied weather, Morse Code, voice procedures and maintenance of the trainer. Students were to attend Control Tower Operators School also located in Atlanta. They flew with pilots on occasion at future duty stations where they learned how to really trust cockpit instruments. The pilot would pull a curtain over the front windscreen and fly by only instruments up to near-touchdown. Pat reported that she knows where the term "white knuckles" came from. Pat Getchell was promoted to lead petty officer at Atlanta with 18 WAVES reporting to her. Life in Atlanta was very good, she reported to me at our lunch get-together. She even learned to love southern fried chicken served at "Mammy's Shanty" in Buckhead. She was, in 1944, transferred to Pearl Harbor, Hawaii and later moved to NAS Kaneohe where she instructed carrier pilots. Pat Getchell was promoted to Chief Petty Officer after only three years of service. She was the first WAVE Chief on the base. As part of the initiation in making Chief, she was thrown off the seaplane dock fully clothed. She vividly remembers that ceremony. She is a remarkable woman who, after training hundreds of pilots in World War II, continues to serve our great country and its citizens

Figure 21—NCS Tribute to WAVE Pat Getchell (H.Okraski)

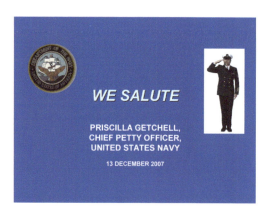

Figure 22—Link Trainer with Operator (US Navy)

Figure 23—Link Trainers with WAVE Instructor/Operators (US Navy)

Figure 24—Logo of the Link Instrument Training Instructors School (US Navy)

Figure 25—An Instructor/Operator Studying Simulator Flight Path (US Navy)

Figure 26—Simulator Plots (US Navy)

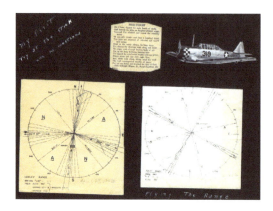

Figure 27 —Ed Link with WAVE Instructor/Operator

Figure 28 —WAVES Report to Duty (Pat Getchell lower far right)
(US Navy)

Special Devices Artificer (SAD)

In 1943 during WWII the Special Devices Artificer was established to gather all of the specialist personnel who operated, instructed on trainers, or repaired and maintained trainers (Special Devices). The following are some of the Special Artificer (Special Devices):

SAD Rating Badge

- Specialist T (LT) Sp(T)(LT) Link Trainer Instructor
- Specialist X (VA) Sp(X)(VA) Visual training aids
- Special Artificer D SAD Knowledge and repair of special training devices
- Specialist G Sp(G) Aviation free gunnery instructor
- SAD(MG)—Special Artificer (Special Devices) (Machine Gun Trainer)
- Sp(A)—Specialist (Physical Training Instructor)
- Sp(G)(N)—Specialist (Anti-Aircraft Gunnery Instructor)
- Sp(T)—Specialist (Teacher)
- Sp(X)—Specialist (not elsewhere classified)

These Sailors and Waves were dedicated team members who realized how important their performance was to the war effort.

TRADEVMAN Rating

In 1948, the Training Device Man (TRADEVMAN, TD) rating was established to replace the SAD rating. The TD Rating was an Aviation rating for Sailors who operated, maintained, repaired, and performed whatever else was needed to make trainers available for the training of Navy trainees. The TD rating was included

TD Rating Badge

in the Aviation Rating Group because most of the existing trainers were aviation trainers. The TDs were the people who "ran" the Link Trainers in the "Link Shack." On major shore Aviation bases TDs were also the keepers of the Film Library and Physiological Training Devices. However, some TDs also served with the Surface and Sub-Surface elements of the Navy, usually in support of Ships Systems Training Devices—Ships Gun Fire Control Trainers—Weapons System Trainers—Submarine Diving Trainers—Steam Propulsion Trainers—Guided Missile Trainers, etc. I had the pleasure of working with Clarence Bachman at the Naval Training Device Center (NTDC) who was one of the original TDs.

TDs were men and women who typically received training in electricity and electronics, hydraulics, pneumatics, theory of flight, aircraft systems, navigation, and operation and maintenance of training devices at TD, Class A School. The school was located at the Naval Aviation Technical Training Center (NATTC), Millington, TN and was 18 weeks in duration. TDs received advanced training after one or two tours in the fleet at TD Class B School which was 22 weeks long. The USMC also sent their Marines to these schools to qualify them for the Training Device MOS.

A historical note: The TD rating was disestablished in 1984 and replaced by Contractor Operation and Maintenance of Simulators (COMS). The COMS program provides contractor operation and maintenance of fielded training systems. COMS typically includes supply support for those systems, and when training system throughput is an issue, may provide facilities support as well. COMS contracts can be structured in many ways such as by: weapon system, site/base, region, or ship class. COMS is typically provided through service-type contracts that are awarded on a competitive basis for an initial period of up to one year, with four additional one year options. The COMS program is coordinated through NAWCTSD.

What was it like to be a TRADEVMAN?

I was fortunate to have associated with several TDs over my career. On the A-7A Weapons System Trainer, I was the Integrated Logistics Manager and two of my team members from the Fleet were Chief Cliff Tully and Chief Mike Phillips, both dedicated and highly skilled individuals. (Figure 29). The success of that program was largely attributable to the Fleet representatives like Cliff and Mike.

I asked Mike to tell what it was like to be a TD. You will find his very interesting story in Appendix A.

Figure 29—Introduction of A-7A WST at NAS Cecil Field 1967
(US Navy)

TD1 M. C. Phillips demonstrates the A-7A Operational Flight Trainer to Captain J. K. Sloatman, Jr., Commanding Officer and Director of the United States Naval Training Device Center, at the A-7A Weapon System Trainer unveiling held last Friday aboard NAS Cecil Field.

Tuskegee Airmen

Getting back to World War II and the people who learned how to fly in the Link Trainer, a group that set their own high water mark was the Tuskegee Airmen. The Tuskegee Airmen were the first African-American military aviators in the United States armed forces. During World War II, African Americans in many U.S. states were still subject to the Jim Crow laws. The American military was racially segregated, as was much of the federal government. The Tuskegee Airmen were subjected to racial discrimination, both within and outside the army. Despite these adversities, they trained and flew with distinction. All black military pilots who trained in the United States trained at Tuskegee, including five Haitians. One thing they had in common was they all had time in the Link Trainer and other simulators including part-task training mock-ups.

It took a considerable amount of convincing and politics to literally get the Tuskegee Airmen "off the ground." Much of their success is due to the efforts of Mary McCloud Bethune and a famous aviatrix Willa Beatrice Brown. Elenor Roosevelt, who actually had one or more flights with a Tuskegee Airman, was a strong supporter of the group and used her influence to have the unit established. (Figure 30)

Figure 30—Famous women who got the Tuskegee Airmen "off the ground" (Wikipedia)

Although the 477th Bombardment Group trained with North American B-25 Mitchell bombers, they never served in combat. The 99th Pursuit Squadron (later, 99th Fighter Squadron) was the first black flying squadron, and the first to deploy overseas (to North Africa in April 1943, and later to Sicily and Italy). The 332nd Fighter Group, which originally included the 100th, 301st, and 302nd Fighter Squadrons, was the first flying group. The group deployed to Italy in early 1944. In June 1944, the 332nd Fighter Group began flying bomber escort missions, and in July 1944, the 99th Fighter Squadron was assigned to the 332nd Fighter Group, which then had four fighter squadrons.

The 99th Fighter Squadron was initially equipped with Curtiss P-40 War-hawks fighter-bomber aircraft. The 332nd Fighter Group and its 100th, 301st, and 302nd Fighter Squadrons were equipped for initial combat missions with Bell P-39 Airacobras (March 1944), later with Republic P-47 Thunderbolts (June–July 1944), and finally with the aircraft with which they became most commonly associated, the North American P-51 Mustang (July 1944). When the pilots of the 332nd Fighter Group painted the tails of their P-47s and later, P-51s, red, the nickname "Red Tails" was coined. (Figures 31 and 32). Bomber crews applied a more effusive "Red-Tail Angels" sobriquet.

The Tuskegee Airmen had a family of simulators for pilot/crew training and maintenance training. (Figures 33-35).

Figure 31—Red Tail Aircraft (Wikipedia)

Figure 32—Red Tails Markings (Wikipedia)

Figure 33—Tuskegee Airman in Link Trainer (Wikipedia)

Figure 34—Tuskegee Aviation Cadets with an aircraft and in the Celestial Navigation Trainer (Wikipedia)

Figure 35—Maintenance personnel learning on a simulator (Wikipedia)

The story of one of those brave airmen is very interesting and inspirational. That person is Charles W.Dryden. (Figure 36).

Figure 36—The Drydens (Wikipedia)

Ret. Lt. Col. **Charles W. Dryden** (September 16, 1920—June 24, 2008) was one of the Tuskegee Airmen. Dryden earned his wings in 1942, and served in the U.S. Army Air Corps in World War II. He is the author of an auto-biography, *A-Train: Memoirs of a Tuskegee Airman.* Dryden was born in New York City in 1920 and graduated from Stuyvesant High School there. He earned degrees in political science from Hofstra University on Long Island and public law from Columbia University in New York City. He was also awarded an honorary doctorate by Hoftstra in 1996. In between, he taught air science at Howard University in Washington, D.C..

Dryden and his comrades performed many duties, it was as bomber escorts that they excelled. By the end of the war, many B-17 and B-24 pilots of the Ninth and Fifteenth air forces began to request "those P-51s with the red tails." My father-in-law, Captain Ralph Dentinger, was assigned to a B-17 unit in England with the Eighth Air Force and held the Red Tails in awe. All totaled, Dryden and his fellow Tuskegee Airmen flew more than 15,000 combat sorties, destroying more than 250 enemy aircraft as well as 950 railcars, trucks, and other motor vehicles. They never lost a single es-corted bomber to enemy action.

But this success came at a high price: sixty-six of Dryden's fellows were killed in action, and thirty-two became prisoners of war. For their efforts the men won 150 Distinguished Flying Crosses (an award first given to Charles Lindbergh), 744 Air Medals, eight Purple Hearts, and fourteen Bronze Stars. Their achievements convinced many white veterans that African Americans were every bit their equals and deserved better treatment back home. As a result, on July 26, 1948, U.S. president Harry S. Truman signed the executive order that desegregated the U.S. military.

Dryden moved to Atlanta in 1996 and that same year received an honorary doctorate of humane letters from Hofstra University. The year 1997 brought much acclaim and attention to Dryden: Georgia secretary of state Max Cleland proclaimed him an "Outstanding Georgia Citizen"; he participated in the opening of the Museum of Aviation display "America's Black Eagles. The Tuskegee Pioneers and Beyond"; and he published his autobiography, *A-Train: Memoirs of a Tuskegee Airman.* The following year he was inducted into the Georgia Aviation Hall of Fame. *(Wikipedia)*

Night Witches

The issue of placing women in combat situations as been around for a long time. In the United States, historically women were not allowed to fight alongside the men. They could ferry aircraft and even be test pilots, serve as nurses and perform other rear echelon duties but combat was out of bounds. However, on January 24, 2013, Secretary of Defense Leon Panetta removed the military's ban on women serving in combat, which was instituted in 1994. Implementation of these rules is ongoing.

The Women's Air Service Pilots (WASP) in the United States fulfilled a very important role of flying support aircraft. The men were engaged in direct combat and could not be spared to do the work the WASPs did. Other countries have used women in combat for many years. One such group was the Soviet Union's "Night Witches", an incredible group of Soviet women who flew bombing missions during World War II. Earlier in this book, I mentioned that Ed Link sold some of the "blue boxes" to Russia. I have to surmise that they were used to train military pilots. I somewhat verified this when I visited many of the training facilities in Russia in 1995. At a major training facility outside of Moscow, I was told by a Russian training expert that he remembered seeing the Link Trainer. However, I will be the first to admit that I have no real proof that the Night Witches used them. I am getting a little ahead of myself here. I need to tell you more about who the "Night Witches" were.

In 1941, Hitler had invaded the Soviet Union. It was a big push that the Russians were not prepared for. By November, the German army was less than 20 miles from Moscow. Leningrad was under siege and 3 million Russians had been taken prisoner. The Soviet air force was grounded. Those were desperate times, requiring desperate action.... enter Marina Raskova.

Marina Raskova, (Figure 37), a record-breaking aviatrix, was called upon to organize a regiment of women pilots to fly night combat missions of harassment bombing. She was selected by Josef Stalin for this position. Marina and two women were the first women to be awarded the Hero of the Soviet Union medal for their record-breaking, 3,240 nautical mile flight. It was Marina's accomplishments and visibility that helped her persuade Stalin to form the three regiments of women combat pilots, aircrew and mechanics. These became the "Night Witches".

Figure 37—Marina Raskova—the Soviet Amelia Earhart

The reason the Germans called this group of female pilots the Night Witches was because the Witches developed a technique of flying close to their intended targets, then cutting their engines-silently they would glide to their targets and release their bombs. Then they would restart their engines and fly away. The first warning the Germans had of an impending raid was the sound of the wind whistling against the wing bracing wires of the Po-2s, and by then it was too late- like "Night Witches" were off into the sky!

There are several books written about this group of female pilots, aircrew and mechanics. The Night Witches had an enviable war record with some pilots earning the "ace" designation. They developed unusual tactics based on the very slow aircraft they flew, taking advantage of their aircraft performance envelope and their own cleverness. The use of decoys in making a low level harassment bombing runs, evading searchlights, was ingenius... but dangerous. Some flew 16-18 missions in a single night nearing exhaustion. (Figure 38). Success in the air spawned wild rumors by the Germans. They offered that the female pilots were injected with some fluids taken from cats, enabling them to see in the dark.

Figure 38—Katya Ryabova and Nadya Popova.18 bombing runs in a single night (www. Seizethesky.com)

(www.SeizetheSky.com—Linda Dowdy)

There are some very interesting stories told about the Night Witches. One of them has to do with a Witch shooting down a German pilot in an air-to-air dogfight. After taking a round of bullets, the German pilot abandoned his burning aircraft via parachute, landed on the ground and was captured by the Russians. After interrogation, the German pilot was asked if he would like to meet the Soviet pilot that shot him down. He offered that any pilot who could out-maneuver him, a superior pilot, must be quite a man. They brought in the slightly built Russian pilot, wearing the sheepskin-lined helmet and goggles of the day, and when the pilot removed that helmet, her hair flowed downward, causing the German pilot to gasp. "What an insult!!" he exclaimed. However, when the Soviet pilot went on to describe the series of events leading to the downing of the plane, the German pilot conceded that ONLY the person who shot him down would have known those maneuvers. It is said that he shook her hand out of respect. *(www.seizethesky.com)*

The Witches trained at Engels Military Aviation School. What follows is some additional information about the school where the they trained. *(Wikipedia)*

By 1936 the Engels military aviation school was one of the best flight schools in the country. Students flew Polikarpov U-2, Polikarpov R-5 and CSS aircraft. Prior to the Second World War the school trained several thousand pilots. Many of them fought in the Spanish Civil War, participated in the Battles of Khalkhin Gol and the Soviet-Finnish War (1939-1940). For participation in the fighting seven pupils were named Heroes of the Soviet Union.

At the beginning of the Second World War the school had in service Polikarpov U-2, ANT-40, Po-2, and others aircraft. During the war years the Engels Flying School sent to the front 14 regiments. Among them were three women's regiments (including the Night Witches). Ninety pupils of the school were awarded the title of Hero of the Soviet Union for combat exploits. *(Wikipedia)*.

We know that Ed Link sold some of his Blue Boxes to Russia but no one seems to know where in Russia they went.. I have to believe that if they were anywhere they would have been at Engels. Connecting the dots, I conclude that the "Night Witches" had used the US-made trainers. That may be a stretch-but I am sticking with it! *(www.SeizetheSky.com—Linda Dowdy)*

Luis de Florez Travels to Britain

In 1941, Commander Luis de Florez (Figure 39) became aware of the initiatives in Britain toward synthetic training technology. Apparently, he was impressed by the developments taking place there. Upon his return, he published a report, "Report on British Synthetic Training." This document established what was to be the future of Navy simulation. One of the simulators he observed while there was the Silloth Trainer. (Figure 40).

The Silloth Trainer was developed by Wing Commander Iles at RAF Silloth, south of Carlisle. The figure shows one of these trainers for a Halifax bomber. The Silloth Trainer was designed for the training of all members of the crew, and was primarily a type of familiarization trainer for learning drills and the handling of malfunctions. As well as the basic flight characteristics, all engine, electric and hydraulic systems were simulated. An instructor's panel, visible in the figure, was provided to enable monitoring of the crew and malfunction insertion. All computation was pneumatic, as in the Link Trainer. Silloth trainers were manufactured for 2 and 4 engine aircraft throughout the war; in mid-1945, 14 of these trainers were in existence or on order. Towards the end of the war a Wellington simulator was developed at RAF St. Athan, using contoured cams to generate the characteristics of the aircraft's flight and engines. This machine, however, did not supplant the Silloth Trainer, as all activity on these ceased at the end of the war.

In the United States, a Silloth Trainer was constructed and evaluated at the Mohler organ plant. (Keep in mind that the Link Trainer was also constructed in the Link Organ Plant). It became obvious that the underlying technology of pneumatics had run its course because it could not offer the fidelity necessary for training and was rather limited in its use. Consequently, Bell Telephone Laboratories was contracted to build an electrical version of the Silloth Trainer. The result was an operational Flight Trainer for the Navy PBM-3 Aircraft. Essentially, this was said by some to be the first flight trainer with specific aircraft flying qualities. *(Rolfe and Staples)*

Figure 39—Commander Luis de Florez

Figure 40—Silloth Trainer

In 1940 Rediffusion, whose manufacturing division later became Redifon, built a direction finding trainer for ground operators. This simulated the Bellini-Tosi goniometer (angle measurement device) and Direction Finder (DF) equipment, whereby two such stations could take a fix on an aircraft transmission and pass the resulting information back to the pilot. A similar trainer was designed to train the operators of Very High Frequency (VHF) stations to give fixes to fighter pilots. However, the most important member of this family of Redifon trainers was the C 100 DF and navigational trainer which was first produced in 1941 to train air crews in the skills of navigation using ground beacons. There is no wonder that Commander de Florez was impressed with the simulator developments in Britain. They were leading the technology of the time. The trainers were installed in five separate cubicles which housed the trainee pilot, navigator and radio operators, and enabled these crews, under the control of an instructor, to carry out navigational exercises, plotting their track from the bearings set up by the instructor. This trainer was similar in principle to the other two Redifon trainers. Suitable decoupling was provided so that up to five receivers and goniometers could be operated from one set of transmitting goniometers enabling the instructor, at the cost of limited flexibility, to teach five crews simultaneously. The transmitting goniometers were mounted on a chart at the position of the beacon stations so that the designated north/south stator coils were aligned with the meridian passing through the particular beacon. The DF receivers were standard RAF airborne units and it was thus possible to tune them in and operate them as would be done in real life. The complete receiving goniometer stators could be physically oriented by the "pilot" of each aircraft to correspond to the aircraft heading during the flight. This was ingenious for that era.

The equipment had provision for the superimposition of interference such as enemy jamming. Some installations were equipped with sound effects and epidiascopes (projections) so that pictures of target areas and other landmarks of importance could be projected in front of the trainer. These installations were known as Crew Procedures Trainers. Well over 100 of the C 100 navigational trainers were built and installed on RAF Bomber Command operational training units and navigational training stations throughout the country and in Canada at the Empire Air Training Stations until the end of the war, plus the small number of trainers installed on USAF stations in this country.

In late 1942, Rediffusion was instructed to install this equipment on the first of the American 8th Air Force's stations at Bovingdon, which was known as a crew replacement centre. (If you recall, my father-in-law was assigned to the 8th Air Force in B-17s.) The American authorities quickly appreciated the benefits of this trainer and requested that it be made to operate with American equipment as installed in the B17 Flying Fortress. In 1943, Rediffusion developed for the American Air Force a Dead Reckoning Navigational Trainer to train up to ten navigators flying in formation. The production model of this trainer, the C500, utilized the C100 and provided hyperbolic Gee fixes with an existing static Gee trainer. (Gee was the code name given to a radio navigation system used by the RAF in WW II) *(Wikipedia)*

One of the best technological successes of the war was the part played by the Trainer Group at the Telecommunications Research Establishment (TRE) in the design of synthetic radar trainers. This group, under G.W.A. Dummer, developed trainers for all of the new radars developed during the war years. In addition to devices attached to Link Trainers, a novel flight simulator for training in AI (Aircraft Interception) was invented. This trainer, the Type 19, was a complete crew, fixed base, trainer for AI combat, which consists of four stages: following an interception course provided by a ground operator, guided by on-board radar, visual contact and the moment of firing. (Today, we refer to AI as Artificial Intelligence-no connection, however). Type 19 provided training in the complete sequence by provision of positions for the pilot and AI operator, and instructors unit, computers for simulation of the attacking aircraft and the relative position of the "enemy", a visual projection unit and a course recorder. The flight simulation computer (known as the Type 8, Part II) was used in a number of TRE trainers, including mobile units whose function was to tour operational squadrons to train in the use of the latest versions of airborne radar. The visual projection system, designed by A.M. Uttley, was used in the larger AI training installations at RAF Operational Training Units. The image, displayed on a hemispherical cyclorama mounted in front of the pilot, consisted of a night sky and ground of controllable brightness with a tail silhouette of a bomber which moved correctly in bank, range, azimuth and elevation in response to relative movements of fighter and bomber. The first AI crew trainers went into service in 1941, while the first complete Type 19 trainer was installed in 1943. It has been estimated that the use of the TRE synthetic radar trainers saved £50,000,000 worth of aviation fuel alone.

In addition to the trainers mentioned, above many others were developed by adding extra features to the basic Link Trainer for such tasks as gunnery instruction. In Britain, the JVW Corporation Limited, formed to market and service Link Trainers, successfully produced a torpedo attack trainer for the Royal Navy, a tank trainer for the Army, and a night vision tester and glider station keeping device for the RAF. The epidiascope visual system for the Torpedo Attack Teacher was produced by Strand Electric, better known for stage lighting. Another simulator with a strong visual element was the Royal Aircraft Establishment's Fixed-Gun Trainer for fighter pilots, developed towards the end of the war, the needs for training in more specialized skills were met by the adoption of a multitude of purpose-built devices. (www.ntlworld.com)

Curtiss-Wright

Although Ed Link produced the first aviation training device, his design did not replicate the true aerodynamic functions of an actual aircraft. Dr. Richard Dehmel of Curtiss-Wright came along and simulated, through analog computation, the actual equations of motion for an aircraft, with the instruments and controls responding just as they would in an aircraft. His patent described the "flight duplicator" that was to go on and give the Link folks a run for their money. (Figure 41). To lighten the competition initially, the flight duplicators simulated heavy aircraft like the B-25. The trainers were easily recognized because they looked like bright, miniature Airstream Trailers. (Figures 42-44) In my field assignments with Link Aviation, I met several Curtiss-Wright Tech Reps supporting the air force on bases throughout the country and overseas. Although there was competition between the two companies, we helped each other when there were difficult problems to solve. Our common objective was to maintain trainer availability so that pilots can have a trainer that is fully operational when they need to train.

Figure 41—Curtiss-Wright Analog Trainer (Wikipedia)

Figure 42—Curtiss-Wright Dehmel Flight Duplicator (Wikipedia)

Figure 43—B-25 Dehmel Flight Duplicator Cockpit
(Wings of Eagles Museum)

Figure 44—Curtiss-Wright Instructor Console
(Wings of Eagles Museum)

Lockheed-Burbank

Simulation can take many forms. An unusual drama took place during World War II at the Lockheed-Burbank aircraft plant on the west coast of the United States.

As a young boy, I recollect the air raid drills we had and the emphasis on preparing for enemy attacks. When a centrally located siren sounded in the neighborhood, we were required to go indoors and turn off the lights. Civil Defense personnel would patrol the neighborhood, looking for violators. Our coastlines were especially vulnerable with the possibility of the Japanese attacking the west coast and the Germans the east coast. Lockheed-Burbank adopted a novel approach to cover and concealment. The entire facility was covered with camouflage that looked like a country scene-nothing the attacking Japanese crews would be interested in! (Figures 45-47)

Figure 45—Lockheed-Burbank in the open

Figure 46—Lockheed Burbank under cover

Figure 47—Lockheed Burbank views under and over the covering

Luis de Florez and Port Washington (Sands Point)

During WW II, Luis de Florez set up his newly-formed special devices organization in a Chevrolet garage in Washington, DC. (Figure 48). He was on the hunt for a new location when he discovered that the old Guggenheim estate at Sands Point, Long Island, was vacant. (Sands Point is adjacent to Port Washington, New York on Long Island). It had been donated by the Guggenheims to the Institute of Aeronautical Sciences. Gugenheim was very interested in aviation and was a close friend of Charles Lindbergh. That is another story in itself, saved for another time. Anyway, Luis de Florez seized the opportunity and moved his unit to the north shore of Long Island.

Figure 48—World War II Washington, DC Home of Special Devices
(US Navy Photos)

Howard Gould, son of railroad tycoon Jay Gould, began construction of the estate after purchasing the land in 1900. Initially, the plan was to build a castle that was to be a replica of Kilkenny Castle. Castle Gould, as it came to be called, was intended to be used as the main house. However, the Goulds did not like the castle so they decided to create another house on the estate which would serve as the main dwelling.

After the completion of this house in 1912, the Goulds sold the estate to Daniel Guggenheim. Upon buying the estate, the name of the main house was changed to Hempstead House (the limestone stables and the servants quarters are, today, still referred to as Castle Gould). In 1917, the Guggenheims donated the estate to the Institute of Aeronautical Sciences. Soon after acquiring the estate, the institute sold it to the U.S. Navy who held it from 1946-1967. The U.S. government declared the estate as surplus and eventually gave the deed of the property to Nassau County, New York in 1971. *(Wikipedia)*

It was picture perfect, a replica of a European castle. There were two main buildings: an engineering building (which was the horse stable when it was occupied by Gould) in the background of Figure 49 and an administration building (this was the resident's living quarters) in the foreground. A note of interest: at the close of World War II, several German scientists were relocated to the United States. It was called "Operation Paperclip". The Center had one or more of those scientists. Their work was classified but folks surmised it was to continue rocket research. When I came on board in 1962, the operation was in full swing. The Technical Director soon became Dr. Hans Wolff, himself a scientist who was freed from a concentration facility in Germany at the close of the war. He shared some of his experiences with us. A role that he played in the German camp was that of electrician. He said that he would occasionally cause a circuit to fail, causing havoc with the German guards. "Get Hans!", they would shout. Hans would correct the problem and they were pleased with his performance. He said that dependence on him and his "skill set" kept him alive.

Our next door neighbor to the "castle" was Mr. Perry Como, a fantastic singer of the Sinatra era who at one time was a barber. On occasion, we would see him strolling along on his grounds, swinging his golf club at imaginary little white balls. He elevated relaxation to a new level. We were to leave that "paradise" beginning in 1965 for what was to be another "magical" area, Orlando, Florida.

Figure 49—Castle Gould, Home of NTDC 1946-1967

The Naval Training Device Center (NTDC) had some unusual research and development projects. One of them was an Atomic Bomb Simulator. (Figure 50). It was a 165 pound bomb of the high explosive incendiary type, and not radioactive. It was designed to be dropped from an aircraft. The thinking was that we needed to demonstrate what the "mushroom cloud" of an atomic bomb detonation was supposed to look like. So, a miniature atom bomb was developed. It was "detonated" one day but no one bothered to let the local authorities know that there would be a big bang and mushroom cloud! Fire engines, police cars and other first-responders paid us a visit. It took a while to describe what had taken place.

Figure 50—Simulated Atomic Bomb (US Navy Photo)

Note: Probably not known by today's employees at the Naval Air War-fare Center Training Systems Division in Orlando is that the overhead projector and the carrousel projectors were developed by NTDC in Port Washington, NY. The overhead projector was originally used to teach navigation where an instructor could write with a grease pencil on a transparency for the class to see. The carrousel projector was developed to teach aircraft target recognition. Early slide projectors used linear slide trays but the Navy wanted a continuous stream of targets (aircraft silhouettes) to be displayed, and they wanted to vary the amount of time each target was to be shown to the students. These two devices, devel-oped under Navy contracts, had tremendous applications to the com-mercial world, as we all know.

Chapter VI

SPACE PROGRAM SIMULATORS

Knowledge and training for the new frontier
using simulation...there was no alternative

Without simulation we would not have a space program. The National Aeronautics and Space Administration (NASA) and its supporting research centers and universities use simulation to analyze all aspects of space travel and the space environment. Simulators, from the outset, were used to train astronauts and the support crews in all phases of space exploration. One cannot practice going into space by going into space. Simulators offer the platforms and environments that allow our brave men and women to practice.

My first personal experience with NASA was in 1964 when I was with the Naval Training Device Center at Port Washington, NY and was being approached by NASA recruiters. They were asked not to recruit Department of Defense engineers and other specialized personnel. Instead, NASA recruiters used means other than coming in the front door to talk to our engineers. At his invitation and after normal working hours, I met with a NASA recruiter at a diner in Roslyn on Long Island.

NASA engineers were "gods "in those days, wearing the white lab coats with the NASA logo and people hanging on every word a NASA person uttered. I recall attending a conference on reliability and maintainability in Los Angeles and listened in awe to Dr. Werner vonBraun describe NASA's future plans. Those were exciting times and the country was totally on board, wanting to move ahead of the Soviet Union. Sputnik was a wake- up call for the United States. I wanted to be a player in the space program. However, I was told that if I went with NASA I would be moving to Houston, Texas after a short period at Grumman Aircraft on Long Island. Having just bought a new home and with a promising career with the Navy, we decided to stay with the device center. That was a very good decision- now that I look back into the rear view mirror of time.

What was so great about the space program was that NASA had not only the commitment from the President but Congress and the people of the United States of America to do something big in space. President Kennedy set the objective of putting a man on the moon and we were off and running. To meet this objective, and learning from our experiences with simulation for the military, NASA set requirements for a vast array of simulators to facilitate space exploration. Most of the human astronaut training was to be done at Houston, Texas. However, there would be simulators to train and conduct practice for the launch teams at Cape Canaveral. (Figure 51).

Figure 51—Kennedy Space Center Firing Room Simulator (Wikipedia)

In Firing Room 1 at Kennedy Space Center (KSC), Shuttle launch team members put the Shuttle system through an integrated simulation. The control room was set up with software used to simulate flight and ground systems in the launch configuration. A Simulation Team, comprised of KSC engineers, introduce 12 or more major problems to prepare the launch team for worst-case scenarios. Such tests and simulations kept the Shuttle launch team sharp and ready for liftoff. *(Wikipedia)*

Project Mercury Mission Simulators

When the time came to design the Mercury flight simulators, experience with aircraft simulators and with those built for the X-15 rocket plane were all that were available. There is one critical difference between training needs for test pilots of aircraft and those of astronauts. Although flying experimental aircraft is always dangerous they are rarely taken to their projected limits on the first flight. Even the X-15 had a long series of buildup missions, first with a smaller engine, later incrementally increasing speed, then altitude, until a series of full out flights sent the plane to the edge of space. In rocket flight the spacecraft is pushed to the outer limits of stress and endurance from the instant of ignition. Its crews must be fully prepared for all contingencies before the first flight and continue to be prepared for every flight afterwards. There are no live "rehearsals".

The primary simulator for the first manned spacecraft was the Mercury Procedures Simulator (MPS), of which two existed. One was at Langley Space
Flight Center, and the other at the Mission Control Center at Cape Canaveral. Analog computers calculated the equations of motion for these simulators, providing signals for the cockpit displays.in addition to this primary
trainer, a centrifuge at the U.S. Naval Air Development Center (NADC)
in Johnsville, Pennsylvania, served as a moving-base simulator. A Mercury
capsule mock-up mounted at the end of the centrifuge arm provided ascent
and entry training. Additionally, Langley built a free-attitude trainer that
simulated the attitude control capabilities of the spacecraft and two part-
task trainers for retrofire and entry practice.

Analog computers commonly supported simulation in the 1950s and early
1960s. Having the advantage of great speed, the electronic analog computer
fit well into the then analog world of the aircraft cockpit and its displays.
By 1961, though, it became obvious that the simulation of a complete or-
bital mission would be impossible using only analog techniques. The types
and number of inputs and calculations stretched the capabilities of such machines so that when NASA defined requirements for Gemini simulators,
digital computers dominated the design. *(NASA History.nasa.gov/computers)*

Gemini Mission Simulators

Training crews for the more complicated Gemini spacecraft and its propor-
tionately more complicated missions required the use of digital computers
in the simulators. Aside from the tasks done during Mercury, such as ascent,
attitude control, and entry, the Gemini project added rendezvous and controlled entries utilizing the spacecraft's greater maneuvering capabilities. At
the Manned Spacecraft Center, NASA installed simulators to provide training for these maneuvers, including a moving-base simulator for formation
flying and docking and a second moving-base simulator for launch, aborts,
and entry. Besides these, two copies of the primary Gemini Mission Simulator, which had the same purpose as the Mercury Procedures Simulator, and

the Johnsville centrifuge completed the list of Gemini trainers. One of the Mission Simulators was at Cape Canaveral; the other at Houston.

Gemini Mission Simulators used between 1963 and 1966 operated on a mix of analog and digital data and thus are a transition between the nearly all analog Mercury equipment and the nearly all digital Apollo and later equipment. Three DDP-224 digital computers dominated the data processing tasks in the Mission Simulator. Built by Computer Control Corporation, which was later acquired by Honeywell Corporation, the three computers provided the simulator with display signals, a functional simulation of the activities of the onboard computer, and signals to control the scene generators. The same computer was being used in flight simulators that we were building for the Navy.

Functional simulation of various components was made easier by the use of digital computers. In a functional simulation, the actual component is not actually located in the simulator, its activities and outputs being created by software within the computer. Thus, in the Gemini Simulator, the on-board computer was not installed, but the algorithms used in its programs were resident in the DDP computers, and when executed, activated computer displays such as the incremental velocity indicator just as on the real spacecraft. From these experiences, we learned that part of the design systems engineering it was necessary to weigh the alternatives of simulation vs. stimulation. One of the issues in using the on-board computer in a simulator is the inability to "freeze" the exercise-stop it where it is in the program which may be necessary for training.

Scene depiction in the Gemini era still depended on the use of television cameras and fake "spacescapes", as in aircraft simulators. Models or large photographs of the earth from space provided scenes that were picked up by a television camera on a moving mount. Signals from the computers moved the camera, thus changing the scene visible from the spacecraft "windows," actually Cathode Ray Tubes (CRTs). A planetarium type of projection was also used on one of the moving-base simulators at Johnson Space Center to project stars, horizon, and target vehicles.

Gemini simulations often included the Mission Control Center and world-wide tracking network. No commercially available computer could keep up with the data flowing to and from the network during these integrated simulations, so NASA asked the General Precision group of the Link Division of Singer Corporation to construct a special-purpose computer as an interface. Singer held the contract for the simulators under the direction of prime contractor McDonnell-Douglas, which supplied cabin and instrumentation mock-ups. Fully functional simulators came on line at the Cape and Houston during 1964.

Moving-base simulation came into its own during the Gemini program. The docking simulator was in a large rectangular cube that permitted great freedom of motion in training crews for station keeping and docking. The dynamic crew procedures simulator that replicated launch, abort, rendezvous, tethered (with the Agena upper stage), and entry maneuvers and procedures suggested the feeling of acceleration at lift-off by tilting the spacecraft at a rate equal to the g buildup during launch from about a 45-degree angle to nearly horizontal to the floor. This resulted in a push on the astronaut's back, which increased from 0.707g to 1g. Engine cutoff and weightless flight could be suggested by returning the spacecraft to its original position, giving a feeling of maximum comfort to the crew. Negative gs could be simulated by tilting the nose down, causing the astronauts to feel their weight on their shoulder harnesses.

Apollo Simulators

Designing and using the Gemini simulators gave NASA considerable experience in producing high fidelity simulations. Actual flight experiences from Mercury went into improving the Gemini simulators. Gemini rendezvous and maneuver experience helped make the Apollo simulations better. NASA adopted some of the Gemini equipment for Apollo. The use of

Honeywell's DDP-224 computers continued, while moving-base simulators were adapted to Apollo use by changing the spacecraft mock-up and modifying existing techniques. Still, the Apollo program requirements demanded a further increase in the amount of computer power. (Figures 52-55).

Figure 52—The Apollo Command Module Mission Simulator.
(NASA photo 108-KSC-67PC-178)

Figure 53—Astronaut Neil Armstrong in Mission Simulator
(Wikipedia)

Figure 54—A shuttle astronaut training in the Neutral Buoyancy Laboratory. (Wikipedia)

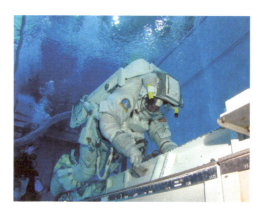

Figure 55—Apollo 11 astronauts Neil Armstrong (left) and Buzz Aldrin train in Building 9 on April 18, 1969.

A "tongue-in-cheek" fact in the movie "Apollo 13" is that the real hero of the mission was the simulator (not Tom Hanks). If you recall, when the crew was having difficulty in determining the optimum electrical load, the team on the ground went to the simulator to try various circuit schemes and they came up with the right loading. In a life-threatening situation as that the fidelity of the simulation must be exact!

NASA Full Fuselage Trainer (FFT)

The FFT is a full-scale mockup of the space shuttle orbiter—without the wings. It was used as a test bed for upgrades to the shuttle fleet and for astronaut training such as extra-vehicular activity (EVA) and emergency egress. Built at Johnson Space Center in the 1970s, was the oldest mockup in the Space Vehicle Mockup Facility (SVMF). The FFT includes flight-quality systems, such as a payload bay, lighting and closed circuit TV (CCTV).

The Space Vehicle Mockup Facility (SVMF) was located inside Building 9 of Johnson Space Center in Houston, Texas. It housed several space shuttle mockups, including the FFT, as well as mockups of every major pressurized module on the International Space Station. It was primarily used for astronaut training and systems familiarization.(Figure 56).

It typically took at least a year and sometimes longer for astronauts to train, depending on the objectives of the mission. Each crew spent up to 100 hours training in the SVMF in more than 20 separate classes.

While many of the systems in the SVMF are flight-like, they do not contain what are generally known as simulators (as used to train pilots). Instead, the FFT and other trainers in the SVMF were used for astronaut training in housekeeping, in-flight maintenance, stowage familiarity, ingress/egress, etc.

It took a versatile team comprising a variety of skills and experience to develop, maintain and operate the Space Vehicle Mockup Facility. Specialists such as designers, engineers, project managers, electronic technicians and shop technicians were used to create the accurate mockups to train astronauts, test systems and procedures, and serve as gravity-bound simulations. *(Museum of Flight)*

Figure 56—Space Shuttle Simulator Cockpit
(Museum of Flight)

A dramatic photo of the simulated Hubble Telescope being released from the space shuttle is shown in Figure 57.

Figure 57—Simulated Hubble Telescope out Shuttle Bay
(NASA Photo)

Personal Experiences with NASA

In 1993, when I was with the Navy as the Director of Research, a team of industry, academia and government members united to submit a proposal for technology transfer of defense technology to civilian use. Janet Weisenford was the lead for the Navy. Janet reported to me. The forerunner of the National Center for Simulation (NCS) was the Training and Simulation-

Technology Consortium (TSTC). The organization was born with NASA assuming organizational responsibility for project management. Ms Priscilla Elfrey was the NASA team leader, providing guidance as the fledgling organization took wings. She continues to be very active in simulation and Science, Technology, Engineering and Mathematics (STEM) initiatives. We address the National Center for Simulation (NCS) later in this book.

Another interesting and challenging experience with NASA for me was when serving as the Vice President for Research of a small company just outside of Orlando where I managed a Small Business Innovative Research (SBIR) project. The objective of the project was to be able to model and visualize possible fracture points of the Next Generation Space Telescope (NGST) when it undergoes stresses. We used Finite Element Analysis (FEA) to create the scientific model. I had the pleasure of working with Dr. Dave Dryer who specializes in data visualization. The project made it through phase 2 and was successful from all aspects. (Figure 58). We were also working on another project that had to do with the adverse effects of substance abuse on children using virtual reality technology-but that is another story. Unfortunately, the company fell on hard times and we had to move on. I became a simulation consultant and still function in that role.

Figure 58—Finite Element Analysis Visualization of NGST

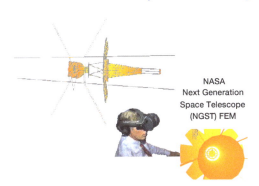

NASA
Next Generation
Space Telescope
(NGST) FEM

Chapter VII

MAKING OF A SIMULATION ENGINEER

Welcome to the Top of the World and the
beginning of a marvelous career..

Entering the Simulation Workforce

After graduating from Clarkson College (now Clarkson University), I took a
position as a Field Engineer with Link Aviation in Binghamton, New York,
following the footsteps of my long-time friend from grammar school, Phil
Wisniewski. I had the pleasure of seeing Ed Link in the Link plant while in
training there. After several weeks of training on their C-11 Jet Instrument
Trainer for the Air Force (Navy calls the same trainer Device 2F-23), I assumed field duties of trainer support in the Southwestern part of the United States. The C-11 was fairly accurate in simulating the Air Force T-33,
Thunderbird aircraft, although that was simply heresay.(Figure 59). I was
assigned to the training installations at Vance Air Force Base (AFB), Enid,
OK, Sheppard AFB, Witchita Falls, TX, Altus AFB, Altus, OK, Tinker AFB,
Oklahoma City, OK, and they threw in, for good measure, Will Rogers Air-

port in Oklahoma City, OK., where they had a C-11 trainer they used for flight research. While at Will Rogers, the Federal Aviation Administration (FAA) was evaluating a Curtiss-Wright commercial flight simulator. Jim Riley, who went on to work with the Navy later, was the TechRep on that first commercial trainer.

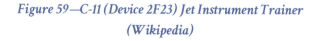

Figure 59—C-11 (Device 2F23) Jet Instrument Trainer (Wikipedia)

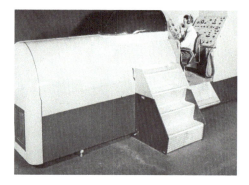

I was very busy making the rounds, performing services, but my schedule was interrupted by a short stint of basic training in the Army at Ft. Chaffee, AR. From there, I went to the F-102A Aircraft simulator school-back to Binghamton, NY. I was given a neat assignment at Suffolk County AFB to work on their simulator at Riverhead, Long Island whrere we were modifying that trainer. That didn't last long... Binghamton was calling.

An Opportunity Not to Refuse

One day in the late summer of 1959, I was summonded to Link Aviation Headquarters in Binghamton, NY to meet with my supervisor, Mr. Ben Previte. Ben sat me down and said, "Hank, do you want to get ahead in this organization?" Oh, oh, I thought, maybe I am in trouble.

"Yes Sir," I replied without hesitation, sliding up to the edge of my seat with anticipation.

"Well, to get ahead, you have to put in some time servicing simulators in a hardship area", he continued.

"OK. What do you have in mind?"

"I can offer you two great opportunities-you can go to Okinawa for 18 months … or you can be assigned to Thule Air Force Base for 6 months."

"I know where Okinawa is. Where is Thule?", I inquired.

He pointed to the top of the world map on his wall that was full of stick pins...but only one where it looks like I might be going. Thule is in Greenland. I looked way up there on the map where all the longitude lines meet, and drew my eyes back to Binghamton. That is a long, long way from here, I thought.

"We need someone to work with the 327th Fighter Interceptor Squadron, maintaining their F-102A Flight Simulator. It is a critical position because the mission is critical. As you may know, the United States has a Ballistic Missile Early Warning System (BMEWS) up there, and it is focused right on the Soviet Union. It is vital that we have aircraft interceptors there to engage enemy aircraft should they decide to cross over the geographic pole. Our Air Force pilots have to be ready, and you are the only person who can provide simulator training." I think he was playing on my patriotism and ego.

With his convincing words, I felt that I was the one who could stop any attack on the United States coming over the north pole. I was already trained on the F-102A and felt comfortable with the systems. I actually liked that simulator...but Thule? Oh well, I can stand on my head for six months. No big deal. I thought it over and, in the mind of a 24 year old, I thought it would be an adventurous and maybe even a fun experience. I agreed to the assignment and was scheduled to leave the following week from McGuire AFB in New Jersey. Little did I know that there was only one flight a week to Thule. Actually, it was a series of flights to Thule. I waited for a week at McGuire until there was a flight to Newfoundland. I was supposed to get arctic clothing at McGuire but the arctic flight scheduler told me that I would be outfitted in Newfoundland when I received a briefing on arctic survival. Had to have that briefing. It was important they told me. Something about the word "survival" that got my attention anyway. So, finally I boarded the C-118 wearing a nice dress suit and tie, with my Link pin displayed proudly on my lapel, heading for Newfouldland. Propeller aircraft were very slow. The world was a lot bigger and slower then. There was no internet or rapid communications. We flew over Goose Bay Labrador and then crossed over to Newfoundland and on to Ernest Harmon AFB where I was to receive "the briefing "and get my arctic gear. I checked in at Operations and requested a briefing from the "survival" officer. I was told that he was on leave-moose hunting. I asked if anyone else could give me what I need to know about the arctic so I could be on my way to my duty station. They needed me there I was told- but the answer was no. Supposedly, he was the only officer qualified to give" the briefing". I spent another week in Newfoundland waiting for the "hunter" to return to the base. Finally, I was ushered into his office where we shook hands and he asked me to take a seat on the other side of his desk.

"I guess you know it gets pretty cold up there at Thule," he said.

"Yes, Sir. Judging from its proximity to the north pole, I figured it got cold up there."

"Do you have any questions," he asked?

"No Sir", I replied. Was this my survival briefing that I waited a week to get?

"I do have one question. Will I be getting my arctic clothing here?"

"No," he said emphaticlly. "You will get it at your next stop in Sondrestrom, Greenland." (now Kangerlussuaq)

"We don't have clothing for civilians here. You understand, don't you? Be careful, it does get cold up there". (I was beginning to get the message: it gets cold in Greenland)

I left quite irritated. I spent two weeks waiting and flown hundreds of miles to be informed that it gets cold in Greenland! When I had geography in third grade at St. Patrick's School we learned about an Eskimo boy who lived in Baffin Island, ate whale blubber, lived in an igloo and paddled his kayak around in search of seals and maybe whales.. I forgot his name but remembered it was cold within the Arctic Circle...for crying out loud!

To add insult to injury, I never was issued arctic gear and went all the way to Thule wearing my nice business suit and tie (my Link pin was beginning to bug me) while other military passengers were bundled up in their fur parkas and arctic pants. We lost most of the aircraft cabin heat at one time and the temperature really dropped. An Air Force first lieutenant felt sorry for me as I sat in a seat made of straps, shivering. He gave me his duffel bag that I went into as far as I could, feet first. I arrived at Thule like a frozen block of ice. My lips were blue, and I was close to hypothermia. I began to understand what Mr. Previte meant by a "hardship" assignment-and I wasn't even there yet! At Thule, I was finally given the clothing necessary to live in that environment. (Figure 60). I arrived in September of 1959, not realizing that in September we were approaching the phase where the curtain of darkness was about to fall and would stay that way until spring.

Figure 60—Author with his Arctic Clothing—finally

That part of Greenland is an arctic desert. There is very little snowfall, the temperature gets down to 50 below zero and the wind can blow with tremendous force, often knocking me off my feet. I recall seeing an empty 50 gallon drum blowing across the road in front of me as I was on my hands and knees trying to make headway through the darkness to the Officers Club. We wore face masks to protect us from frostbite. Without the protection we would be in big trouble, especially if your work brought you outside for long periods of time.

Thule had most of the features of a normal state-side base-a fine gymnasium, bowling alley, chapel, hobby shops, theater, clubs, etc. The food was excellent and cheap. Celebrities would visit around Christmas holidays. Bob Hope was one who visited at one time but not while I was there. Our barracks were comfortable but had an odor due to the manual flush toilets. Stuff kinda backed up. All the techreps: RCA, Hughes, Link, etc. lived in these barracks buildings. I had a room as wide as my arms extended and perhaps twice as long. The building was well heated and comfortable in a spartan manner. Buildings are erected off the ground, otherwise they would sink into the permafrost because of the heat of the structure. We had television-not live, but tapes were flown in on a weekly basis. Everything like news, sports, etc. was a week late. The" uninformed "easily lost wagers on baseball or other sporting events that took place a week before. There was no telephone communications directly with the states. To make a telephone call, you had to contact the base radio operator who would radio a "ham" operator in the states who, in turn, would place a long distance call to your party. Usually, it didn't work. Most folks gave up after a while and resorted to snail mail. There was an Eskimo village nearby but we were not allowed to go there because we were told that non-Eskimos carried so many germs that we could wipe out the entire community. Occasionally, the representative of the Eskimos would slip over to the barracks and sell items made by the natives. Carved walrus tusks were a favorite purchase. Also, occasionally an Eskimo would show up in the base exchange. Not sure if they were there to purchase anything or just look around. When their furs began to thaw, the odor moved through the building at warp speed. Even when they left the building the odor of the furs lingered… a long "hang time".

The 327th Fighter Interceptor Squadron, the "Iron Masks", was activated at George AFB in August 1955. The squadron moved to Thule,Greenland in July 1958 during the Cold War. It was outfitted with F-102A interceptor aircraft. Just a short distance from the base was located a Ballistic Missile Early Warning System (BMEWS). The BMEWS antenna, the size of a football field on its side, was pointed right at the Soviet Union to detect any missiles on their way to the United States. Aircraft would be deployesd if there were an alert. There were plenty of practice alerts while I was there. The F-102A was the first aircraft that could exceed the speed of sound. It had a wonderful weapons system for its time with the MG-10 Fire Control System that could automatically vector the aircraft, via a data link, to an intercept point in space. Pilots could select a pursuit attack or an intercept approach to the target and then fire missiles or rockets. My job and my fellow Link Representative's responsibility was to assist the Air Force personnel in the operation and maintenance of the F-102A Flight Simulator. We were to ensure that the simulator was ready-for-training at all times. We accomplished that objective. A little known fact is that President George W. Bush was an F-102 pilot with simulator experience that he gained while serving in the Texas Air National Guard. That came much later, of course, in the 1970's. (Figure 61).

Figure 61—Lt George W. Bush in the F-102A Aircraft.
(Fiddlers Green)

The primary mission of the F-102 Aircraft (Figure 62) was to intercept and destroy enemy aircraft. It was the world's first supersonic all-weather jet interceptor and the USAF's first operational delta-wing aircraft. The F-102 made its initial flight on Oct. 24, 1953, and became operational with the Air Defense Command in 1956. At the peak of deployment in the late 1950s, F-102s equipped more than 25 ADC squadrons. Convair built 1,000 F-102s, 875 of which were F-102As. The USAF also bought 111 TF-102s as combat trainers with side-by-side seating.

In a wartime situation, after electronic equipment on board the F-102 had located the enemy aircraft, the F-102's radar would guide it into position for attack. At the proper moment, the electronic fire control system would automatically fire the F-102's air-to-air rockets and missiles.

The National Museum of the United States Air Force has an F-102A on display in its Cold War Gallery

Figure 62—The Supersonic F-102A Delta Dagger Aircraft
(Wikipedia)

Let me describe the technology employed in the F-102 Aircraft simulator. The flight simulator (Figure 63) was an analog device utilizing operational amplifiers and several Link Aviation proprietary function generators, control loading and other hardware. The engine was simulated by variable resistors and potentiometers were used throughout the device to simulate various functions in concert with operational amplifiers. There was no visual system attached to the trainer. It was all instruments and controls. Motion simulation was not provided. The benefits of motion in a simulator was (and is) discussed extensively but the customer was not ready to invest in motion unless the benefit in the form of more effective training could be clearly shown. Today, the consensus is that motion is of greatest value in a simulator where there is asymetric thrust (such as with multi- engine aircraft) or where onset cues are important to the pilot (such as helicopters). Ed Link, himself, was a strong advocate of motion. He invested in simulators with motion to demonstrate to customers the realisim that was added through motion cues.

Figure 63—F-102A Simulator Cockpit (Wikipedia)

From a logistics standpoint, the F-102A simulator was a joy to maintain because of its maintenance features: an automatic dc amplifier checking system and a logical network of subsystems. (Figures 64 and 65). Documentation was very good, facilitating a quick repair of equipment failures. We never had a problem with repair parts and, I might add, the factory training Link gave us was excellent. We had 100 % operational availability of the trainer. We were very proud of that. Pilots were never turned away because the system was down.

Figure 64—Maintenance Technician in Greenland checking
amplifiers on F-102A Flight Simulator

Figure 65—Technician in Greenland checking intercept accuracy

The squadron pilots were very dedicated professionals. They practiced radar intercept runs over and over again. They asked us to set up the most difficult intercepts possible. It became a challenge for us to try to "stump the pilot." As a consequence, the 327th FIS pilots became experts in their craft. When they went to compete in the William Tell competition in Florida their performance was outstanding. As they say, practice makes perfect! Of course, there wasn't much to do on off-hours at Thule other than workout, gamble or drink. With total darkness around the clock it could be depressing unless one remained busy. We all spent time under sun lamps to ward off depression. Squadron pilots would practice radar intercepts and emergency procedures in the simulator at every opportunity. There was, indeed, motivation for practicing emergency procedures when you realize that operations took place over the ice of Melville Bay, Baffin Bay or the icecap. A summer picture of Thule AFB is provided as Figure 66. (Winter pictures are not possible)

Figure 66—Thule AFB in the Summer (USAF Photo)

If a pilot had to bail out or set down in the bay, it would most certainly result in death by freezing. Shortly after I arrived at Thule, a helicopter crashed into the bay. The crew and Danish passengers perished quickly in the icey water.

With long periods of boredom, the TechReps found interesting ways to amuse themselves. Two of these characters had an interesting saga in play. The unsuspecting victim of their little drama was the RCA representative. It boils down to this. He bought a tube of toothpaste when he arrived over a year ago and the tube was still close to being full. (Both reps were there for an 18 month assignment. In those days, there would be no income tax for that period. So, engineers and technicians made and saved a considerable amount of money). He would comment on how tightly they pack the toothpaste in those tubes-never had to buy a new one. Remarkable. What he didn't know was that the Hughes representative would, regularly, take his tube of the same brand and squeeze some into the other guy's. This went on up and until the time I left Thule. I often wonder if Mr. Hughes ever was discovered.

After months there on the frozen tundra when the sun began to appear again after a long winter's sleep, I was informed that the 327th was to be deactivated on 25 March 1960, and the simulator was to be moved to Ernest Harmon AFB in Newfoundland. I have to give the squadron commander credit because he had his pilots practice their navigation in the simulator, preparing them for the long flight to Goose Bay, Labrador. This was not a trivial mission not only because of the dangerous environment but magnetic compasses were of no value. In reality, the north magnetic pole was to the southwest of our base. Most people do not realize that the magnetic poles actually move. Gyro compasses were used as navigation aids.

My job in the relocation, as described to me by my Binghamton supervisor via scratchy radio, was to pack and ship the simulator to Newfoundland. The simulator handbook included instructions on shipping the trainer. It began with securing all the cabinets and locking them in place for shipment.

Then we were supposed to crate the larger cabinets in wood. I requested wood from the supply sergeant and received a big belly laugh. There is no wood at Thule, I was told. There are no trees within a thousand miles, he informed me. Anyway, I had my orders. We moved the trainer-cabinet by cabinet with a forklift under the most awful freezing conditions immaginable. It was well below zero and we were out there moving equipment into the hangar for shipment.

I was sent back to the states to England AFB in Louisiana where some of the personnel from the 327th went after the squadron was disestablished. I figured I could now move up the ladder with Link Aviation because I fulfilled my tour at a hardship duty assignment!! My next assignment was to be with the FAA in Atlantic City, but only for a few months.

Welcome Aboard

Once again, Link Aviation gave me an opportunity to excel. It was perhaps the best break of my life. I was assigned to work with the Navy under contract to the Naval Training Device Center (NTDC) in Port Washington, New York. I would not be working in New York, however. I was to participate in what was called the simulator Material Reliability and Integrity Program, operating out of the NTDC Regional Office in Norfolk, Virginia. I worked with two other field engineers inspecting all of the Navy and Marine Corps flight simulators located from Brunswick, Maine to Key West, Florida. Inspection included running a portion of the original acceptance tests and visually inspecting the trainers. The biggest problem with simulators in that era, and to a lesser degree today, is the inability to keep the simulators current with the aircraft simulated. I have to hand it to the Air Force because they contracted Link in those days to monitor changes being proposed to the aircraft and was at the ready to submit an Engineering Change Proposal (ECP) with cost and delivery estimates to incorporate aircraft changes to the simulator. The Navy, at that time, was not that organized nor focused and was less inclined to contract to Link or any other contractor to monitor aircraft change activity. Some of the changes were simple hardware modifications like changing a switch in the cockpit and others might be as complex as replacing the aircraft engine. In any situation, it is imperative that the simulator look and perform like the parent aircraft. To do otherwise, would invite negative training in the simulator. Unfortunately, most simulators were not current and that drove the user community to reject the trainer when it came to using it. Therefore, utilization rates for the family of Navy aircraft simulators were very low.

By our very charter, we had to find and identify trainer discrepancies, short-comings, poor maintenance, etc. These could be a reflection on the maintenance crew's performance. Generally, the trainer crews were courteous and were forthcoming on providing data, etc. Needless to say, if you are a member of any kind of "inspection team" you are not exactly welcomed with open arms. They were not above a little fooling around. One of my duties was to inspect the condition of trailers that housed an aircraft simulator. I recall one hot day in Key West when I climbed a ladder and got on the top of the trailer looking for cracks or "Kool Seal" repairs. While on the roof, someone quietly removed the ladder and left me alone and sweating. Yelling did no good. The crew was in the trailer below, laughing no doubt, and could not hear me. Finally, someone appeared to let me down. After each round of trainer inspections, the Navy commanders from the Atlantic Fleet convened a meeting in Norfolk, VA and discussed the status of each trainer. Actions were prioritized and we were on the way to getting trainers the attention and visibility they deserved.

Along with the configuration issue was the lack of policy on simulator utilization. Up to about the 1970s, utilization was up to the squadron or unit commanding officer. As simulator fidelity improved it became evident that simulator time might be substituted for aircraft training time for certain tasks. Up sprang the need to perform Front End Analyses (FEA) to include the Instructional System Design (ISD) process. Through a task analysis and media selection methodology, the entire simulator's design requirements were formulated and transformed into a simulator specification. The end use of the trainer with curriculum in place determined how the simulator was to be used.

My assignment to the inspection team allowed me to become familiar with the entire inventory of Navy and Marine Corps flight simulators. While in Norfolk, I was sent to several of the fleet training courses such as anti-submarine warfare. These courses were related to the trainers to be inspected. This was a wonderful opportunity for anyone-especially a civilian. In many of the courses, I was the lone civilian there.

I subsequently was hired by NTDC as an engineer in the engineering change group as a GS-11. We moved to Port Washington, thinking that would be a permanent home. No such luck! Three years later, after just purchasing a new home in Northport, NY, the Commanding Officer came on the intercom and announced. "Your new home will be in Orlando, Florida." How can that be, I thought. I have a new home but it is not in Orlando, Florida!

My wife and two children moved to Orlando which was a nice and quiet little town in the orange groves. You could smell the orange blossoms as you drove around on a Sunday afternoon. That move set in motion what was to be the "BIG BANG " of the simulation industry: the beginning of the Center of Excellence for Simulation and Training, recognized by Governor Graham and Senator Bill Nelson who said that the Orlando simulation industry was a "National Asset".

The next twenty-eight years of my Federal Service saw a tremendous growth of military simulation in all of the services and other government agencies. The military services took simulation training much more seriously and policies were put in place to mandate simulator training. The Chief of Naval Operations and the Naval Air Systems Command along with the Fleet and Training Commands put the necessary policies in place. Trainer utilization was tracked and it became an important metric in determining aircraft operational effectiveness. Substitution of simulator time for aircraft or other weapons system time is today a reality. In tough economic times, we need to turn to simulation to maintain our edge over any and all enemies because simulation training is, indeed, a "force multiplier ".

The move to Orlando is a story in itself but because a large number of employees decided to stay in New York, there were opportunities for me to move up the career ladder. I believe I made GS-14 about the time I was 30 years old.

Training Analysis and Evaluation Group

The Navy went through a major restructure of education and training in the 70s. It was at that time the Naval Education and Training Command was formed in Pensacola, Florida. The command was "dual-hatted". That meant that the admiral in charge of training in Pensacola also occupied a seat in the Office of the Chief of Naval Operations (OP 99). In Orlando, a visionary Naval officer, Captain Frank Featherston, created the Training Analysis and Evaluation Group (TAEG). The timing was perfect because it spawned interest in training, developed new requirements and to put emphasis on the Instructional Systems Design (ISD) process. There was no formal infrastructure in TAEG and interdisciplinary teams were formed to address specific issues.

Management was by a "troika" of Dr. Jim Reagan, Paul Little and Jack Armstrong (psychologist, engineer and education specialist-in that order). This was a refreshing undertaking that led to revamping Navy undergraduate pilot training, promoted submarine acoustic training, highlighted the media selection process and introduced modeling and simulation as a means for analysis. I managed a program called "The Design of Training Systems" with the purpose of introducing computer models into the design, planning and execution of training in the Navy. It was a hard sell at that time. This was before personal computers and all of the planning was done by hand. My team of Bill Lindahl, Tom McNaney and Dr. Bill Rankin worked very hard in trying to convince Navy training managers that computer models were the way of the future. Still, heels were dug in by navy senior managers. Many of the changes recommended were not readily accepted by the functional commands. To paraphrase Emerson," whenever you introduce a new order of things, you will have opposition from those who are comfortable in the current system and luke warm support from those who favor the change." Our timing was just not right-but history has proven us to be correct. Time vindicated us! For a period, I was the acting director of TAEG and Dr. Al Smode; a renowned and well-published researcher replaced me. I went back to NTEC as an Engineering division head at that time. NTDC was retitled to the Naval Training Equipment Center (NTEC).

Gratifying Experience

Personally, my career has been a wonderful, gratifying experience. To know that I had a part in saving the lives of countless servicemen and women made it all worthwhile. Yes, I had fun while doing it, with opportunities to ride airplanes, go aboard ships and submarines, making real progress in our laboratories and engineering hundreds of simulators and training systems for the troops.

One does not get things done in a vacuum. It takes a team with a leader to accomplish worthwhile objectives. I define leadership as: "getting people to do what you want them to do because they want to do it." It is that last part that is often disregarded because it is the most difficult. It takes patience and a lot of listening to various views in the process. But once you have commitment, the team is on its way. I was fortunate to have great people working with me when I led the Engineering and Research Departments. (and at one time, the Research and Engineering Department). People like Walt Chambers, Ted Pearson, Paul Little, Joe Rogers, Bill Harris, Bill Parrish, Frank Jamison, Dick Jarvis, Ed Moore, Morris Middleton, Diane Moss, Bill Rizzo, Al Marshall, Carol Denton, Brenda Fulco, Helena Smith. and Jim Bishop were instrumental in making sure we were doing the right thing and doing it properly. We were there to serve the Fleet by delivering tools for effective training.

We ushered-in computer software technology with all of its tentacles of processes, reviews and documentation. The government has to show some leadership and knowledge in letting contracts to industry for software-intensive training systems. Our philosophy was that the government must be a "smart buyer" to make sure the taxpayer is getting their money's worth and we are meeting the user's requirements...and that we did. In our labs we did prototypes and proofs of concepts. I made sure that we had a transition plan for all of our research. In engineering, we had a facility where we did some hands-on systems design and software support. One of the first female project managers, Cathy Matthews, managed the design and development of a Passive Acoustic Analysis Trainer for the submarine community, making the Chief of Naval Operations (OP-29), George Horn, very happy.

I would, and do, encourage young people to consider a career in Science, Technology, Engineering and Mathematics (STEM) but specifically in simulation technology. While with the Navy, we had a very aggressive recruiting program where I visited many colleges and universities in search of graduates who wanted to make simulation a career. We hired some of the best young men and women the system could provide. So many are leaders of government, industry and academia today. That makes one feel real good.

The rewards are great. A short time ago, I was giving a presentation on the "History of Simulation" to the folks at NAWCTSD in Orlando and after it was over a young lady came up to me and said, "You don't remember me.. but you are the reason I am here." Memory racing to recognize the face, I couldn't come up with the time nor place where I knew her from. She helped me by saying that when I visited her local high school several years before, she became interested in simulation and the more she got into it the more she was captivated. That is the kind of experience that raises the hairs on the back of the neck and makes it all worthwhile. NAWCTSD must be a great place to work when you have people like Pete Engel who is the last remaining employed "plankowner". He is one of the first to relocate from Port Washington, N.Y. to Orlando in 1965 and is still actively working for the Navy.

- Chapter VIII -

TO RUSSIA WITH LOVE

A look into the past with the hope of tomorrow...

n June of 1994, after retiring from the government, I was a part of a delegation of 15 computer simulation individuals from the United States along with five others from foreign countries that visited institutes, companies and universities in St. Petersburg, Moscow and Kiev. The delegation was led by Vice Admiral John Disher (Retired) who was then the director of the National Training and Simulation Association (NTSA). The visit was in response to an invitation by the Russian Academy of Science to the Citizen Ambassador Program, a subordinate organization of People to People International. The purpose of the trip was to visit Russia's top organizations engaged in modeling, simulation and training. At that time, Russia had gone through "peristroika" and the defense economy had collapsed. Consequently, Russian industry was seeking new opportunities with the western world and elsewhere. The depression left skilled workers and teachers without jobs. Those that could find work were minimally paid. Teachers drove taxicabs and took other jobs to support their families. Industries were converting from defense to commercial products. Aircraft systems manufacturers were retooling to produce refrigerators and other household appliances. The simulation industry was looking for business or assistance from the United States. People in Russia that we met were very cordial, cooperative and did not hesitate to open their doors to discussions about their products, services and capabilities. They viewed us, I believe, as a possible path out of their economy's slow-down.

Our visits included Leninitz Holding Company in St. Petersburg which includes 35 research institutes, factories and other holdings that have 43,000 employees. They had made the conversion from 90% military work in 1992 to 25% in 1994.

Next was the St. Petersburg University of Marine Technology, a leading university in naval engineering and offers acoustic modeling of sound propagation. Zhukovsky was the home of the Central Aerohydrodynamics Institute (TsAGI) and the Flight Research Institute located about 25 miles southeast of Moscow. TsAGI does fundamental research of flight and reconstruction of flight information in accident investigations. They have some low fidelity simulations.

The majority of the simulators were at the Gromov Flight Research Institute where test pilots are trained for a wide array of Russian aircraft. The facility has simulators for the following aircraft: TU-144, SU-17, SU-17, MIL-8 and MIG-23. These simulators, at the time, were of the 1960 era technology. Their training programs were, however, of high quality. They seemed to make the best out of the assets they could afford.

It was at Gromov where I was asked if I would like to fly the MIG-23 simulator. I was honored and humbled by their invitation. Having flown many simulators in my position with the Navy, I gladly accepted. Climbing into the cockpit, is when it hit me. Suppose, I thought, when they put up a visual aircraft target for me to fly against it would have United States markings on the simulated aircraft? What would I do? This would be very em-barassing for me, my fellow delegation members and the State Department. Fortunately, the hosts were one step ahead and put up a "generic" target aircraft. The rest was fun! Later in the evening after a few shots of vodka (three ounce variety), I expressed my fear and got a big laugh from the hosts. Incidently, toasts are made by our Russian friends with vodka-filled water glasses for just about any cause, and the drinks are downed with reckless abandon. After a couple of those, I turned my attention to a nearby potted plant-if you know what I mean.

We also visited the Moscow Aviation Institute (MAI) which consists of 10 colleges with a faculty of 2,000 in support of 15,000 students. That is a very desireable teacher/student ratio. At MAI we were shown actual aircraft: YAK-42, SU-27, MIG-25, MI-25 and MI-1, along with some simulators. (Figure 67).

Figure 67—The TL39 3-DoF motion simulator with IOS
at MAI University

The delegation traveled to Kiev by railroad. We were told to lock our cab-ins at night because we could be robbed if we didn't take precautions. They explained that Russia has their own versions of Jesse James. As we traveled through the wheat belt of Russia and the Ukraine, I couldn't help but be taken in by the vast expanse of the waves of grain—as far as the eye could see. Then, occasionally, I would see a small building with a farmer with a horse-drawn piece of machinery cutting the grain. How in the world do they cut all that grain with such outdated means, I wondered.

At Kiev, we were at the Kiev International University of Civil Aviation. The institution was very clean and orderly, and impressive. It is the leading civil aviation university in the Ukraine.

Our last stop in Kiev was at the V.M. Glushkov Institute of Cybernetics. It is part of the Ukraine Academy of Science. It develops computer theories, methods and databases, control systems and information technologies. They are working on computer models for aircraft collision, robotics, decision support systems, air traffic control and energy management.

All-in-all, we were impressed with the expertise of our host countries' engineers, scientists and teachers. When their economy improves, the experts will be able to afford more facilities and technologies to improve their overall capabilities.

We had discussions about teaching computers in the schools. Our hosts made an interesting observation, comparing the schools of the United States and Russia. In Russia, the teachers are totally in control in the classroom and are well respected in their social system. When they come to class to teach computers they lead the class because the teacher is more computer literate than any student. This is because (remember this is 1994) the students do not have computers at home. Nor do they have access to them elsewhere. In the United States, teachers are poorly paid and in most classes there are students who know more than the teachers about computers. This is because of the free market system of capitalisim with computers available on the open market.

Walter Ulrich of Halldale Publishing, and an expert on the history of simulation, has current information on the changing Russian simulation industry and he reports:

"Russia's domestic aircraft industry is facing a bright future," said Prime
Minister Vladimir Putin during his annual TV and radio message in
December 2009. At the same time, he announced that the state-owned
United Aircraft Corporation would receive additional financial support
and have its 1.6 billion US$ debt restructured—a strong indication that
the Russian government is still willing to spend to ensure the survival
of the national aircraft industry despite the 2009 criticisms of serious
management errors such as a lack of willingness to invest in new manufacturing plants or the suppression of uncomfortable initiatives. S&T
manufacturers did not escape blame. For far too long they had stuck
to outdated standards that discouraged development of modern simulators. Even if the Russian S&T providers today have products that are
competitive at home and on some export markets, when it comes to
competitiveness in the upper class range, Russian S&T providers still
depend on cooperation with Western companies. And most experts
agree that this is not about to change." *(Ulrich)*

Chapter IX

SIMULATION TECHNOLOGY AND APPLICATIONS

Without the digital computer, simulation technology would be on "freeze"..

Since Ed Link introduced his Blue Box in World War II, the change in simulation technology has been meteoric in every dimension. We owe this primarily to the rapid growth of digital computing systems and a likewise expansion of the commercial market for computing and display systems.

Visual Systems-Model Boards

Rapid advancements have been made in visual technology. In the 60s we experimented with black and white television cameras and moving belts to simulate aircraft landings. (Figure 68). While at the FAA facility in Atlantic City, NJ for a short assignment, I observed experiments involving simulated aircraft landings using the belt and TV system.

Figure 68—Belt-Driven Visual System (AIAA Rpt 93-3547)

Figure 5. Basic Simulator with External Horizon

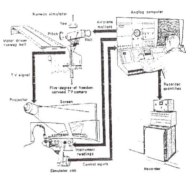

equipped with a transport-type cockpit, a conventional cockpit instrument display, normal flight controls (with control forces provided by springs and dampers), and a general purpose analog computer programmed with math model equations of six degrees of motion freedom.

The simulator was used in a study of the effect of lateral-directional control coupling in the control-system for supersonic and hypersonic aircraft (ref. 10).

Lessons Learned

• Although the wide field of view provided a much improved bank angle reference, the simulator was

Following that somewhat "Rube Goldberg" technology and with improvements in television systems, came full model boards. Model boards were actual replications of buildings, roads, bridges, airfields and other terrain features. (Figure 69)

Figure 69—Terrain Model Board (1959)

Model boards could be mounted horizontally or vertically. A large bank of powerful lights was necessary to provide the illumination needed to give a bright enough presentation to the pilot. A television probe moved along the terrain according to the ground speed of the simulated aircraft. When I moved to Orlando, Florida in 1965 with the Navy, Martin-Marietta had such a system and they were using it for helicopter training research. The problems with the model boards, which were made in England for the most part, were that they were too expensive to build and operate with those powerful lamps, limited in geographic area to be simulated and because the probe had to follow the nap of the earth, the probe would often hit a building model or something and get damaged. The emergence of computer image generation was the prescription for replacing the antiquated model boards. There were only a couple of companies investing in image generation at the time. General Electric was one and Evans and Sutherland, headquartered in Salt Lake City, Utah was another.

The submarine community used model board technology for periscope training before computer image generation of targets and environments came along. I never did see one of these systems in operation but one of my veteran submariner employees, Dick Kramer, described the Submarine Attack Teacher located at New London to me. (Figure 70). On the top deck of a large building was a simulated ocean made of terrazzo that had a curvature just like the earth's surface. On that simulated ocean were five movable models that looked like enemy, friendly and neutral vessels. The models were of 1200/1 scale. On a lower deck (second) was the conning tower with periscope and controls. Optical lenses and mirrors that presented the targets somewhat realistically to the trainee were embedded in the simulated periscope. On the first floor was the diving trainer that controlled the submarine movement through the ocean space. Crews could practice target recognition, angle on the bow, etc. and conduct attacks. It was an expensive, higher fidelity version of the cardboard and mirror periscopes we had as kids. (Incidentally, I installed a home made periscope just like that using

PVC tubing and mirrors about ten years ago in my grandson's tree fort. Today, he is serving in the Marine Corps).

Figure 70—New London Periscope Simulation (Popular Mechanics)

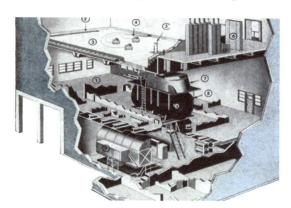

The surface navy, army and marines also used model boards for various training requirements. In fact, sand tables went back a long time and were used by the Romans and others to develop tactics.

I have been told that immediately before the invasion of Normandy in World War II, General Dwight D. Eisenhower met with his senior officers in an orange grove adjacent to the Orlando Air Base in Florida to plan the invasion. (Figure 71). As a "mission rehearsal" aid, he had a concrete scale model of the beachhead of France built and set up in the grove. With miniature models of his assets, he and the officers under his command planned the invasion. This was verified by an "old timer" I met when I was at the Orlando Navy Base in the 60's. He claimed to have a piece of the concrete model. It seems as though, even at that time, Central Florida was predetermined to be a focal point for simulation. (More about that later with a discussion about Team Orlando and the meteoric growth of simulation in Central Florida).

Figure 71—WWII Mission Rehearsal in Florida (Dick Adkins)

I mentioned Dick Kramer, the retired Chief Petty Officer, who spent his career in the submarine force living in those space-limited boats of that era. We are not talking about a "boomer" Trident or the fast attack submarines of today. He was in the diesel/electric boats in World War II where you lived in cramped quarters, getting to know your shipmates real well. He taught me an important lesson on giving "appropriate" feedback to the project manager that I wish to pass on to you.

His boat operated in the North Atlantic where they encountered all types of German surface ships and U-Boats escorting the ships. He revealed some of those experiences to me. A typical scenario is where the boat's skipper would plan and execute an attack strategy on a target ship, maneuver without being detected into an optimum angle and then fire torpedoes. In most cases, they destroyed enemy ship targets, slipping away undetected as submarines are supposed to do. However, there were times, and not all that infrequent, when the torpedo would just "bounce" off the hull of the target ship. Fire another one and again-nothing but a big clanging noise but no detonation. By that time, their boat was detected as they were at periscope depth and, as a result, the destroyers and U-Boats were in hot pursuit. Through the grace of God and the excellent training the crew underwent they were able to escape. However, the story continues. When they arrived in port, a few members of the crew made their way to Washington, DC where they stormed through the doors of the Main Navy Building and asked to see the project manager responsible for the torpedoes used on their boat. Moving down the corridors of the building like wolves after their prey, they found his office, walked in and asked the individual seated at an old metal desk covered with papers if he was the project manager for Torpedo XXX. An affirmative was given and he almost completed the "how can I help..." sentence before the boat crew members grabbed him by both arms, slung him against a bank of lockers and proceeded to give him, feet off the ground, a lecture on how he almost got them killed because of defective torpedoes sent to the fleet. Now, that's what I call constructive feedback from the user! Fear and hope are great motivators. The crew had hopes that he would fix the problem and the project manager was scared that he would get another visit from those guys in the future.

Computer Generated Imagery (CGI)

Computer Generated Imagery (CGI) soon began to replace model boards and everyone knew that was the direction the military services were heading. The first company that I dealt with in that world was Evans and Sutherland, named after its founders. (Figure 72)

Figure 72—Evans and Sutherland Pioneers in CGI (Wikipedia)

Evans and Sutherland (E&S) was founded in 1968 by David Evans and Ivan Sutherland, professors in the Computer Science Department at the University of Utah.

The two professors were pioneers in computer graphics technology. They formed the company to produce hardware to run the systems being developed in the university, working from an abandoned barracks on the university grounds. The company was later located in the University of Utah Research Park. Most of the employees were active or former students, and included Jim Clark, who started Silicon Graphics, Ed Catmull, co-founder of Pixar, and John Warnock of Adobe.

In the 1970s they purchased General Electric's flight simulator division and formed a partnership with Rediffusion Simulation, a UK-based flight simulator company, to design and build digital flight simulators. For the next three decades this was E&S's primary market, delivering display systems with enough brightness to light up a simulator cockpit to daytime light levels. These simulators were used for training in in-flight refueling, carrier landing, AWACS, B-52 EVS, submarine periscope and space station docking. We in the Navy had many meetings with them to discuss what was proprietary in their design. The logistics tail on proprietary items can be very costly on a lfe cycle basis.

In the mid 70s until the end of the 80s, E&S produced the *Picture System* 1, 2 and PS300 series. These unique "calligraphic" (vector) color displays had depth cueing and could draw large wireframe models and manipulate (rotate, shift, zoom) them in real time. They were mainly used in chemistry to visualize large molecules such as enzymes or polynucleotides. The end of the Picture System line came in the late 80s, when raster devices on workstations could render anti-aliased lines faster.

In 1978 the company went public with a listing on the NASDAQ Stock Exchange.

In the 1980s, E&S added a Digital Theater division, supplying all-digital projectors to create immersive mass-audience experiences at planetariums, visitor attractions and similar education and entertainment venues. Digital Theater grew to become a major arm of E&S commercial activity with hundreds of Digistar 1 and 2 systems installed around the world, such as at the Saint Louis Science Center in St. Louis, Missouri.

For a brief period between 1986 and 1989 E&S was also a supercomputer vendor, but their ES-1 was released just as the supercomputer market was drying up in the post-cold war military wind-down. Only a handful of machines were built, most broken up for scrap.

During the 1990s, E&S tried to expand into several other commercial markets. The Freedom Series graphics engine was developed to work with Sun Microsystems, IBM, Hewlett Packard, and DEC workstations. 3D Pro was developed for the first wave of 3D graphics cards for PCs. Also, the MindSet virtual set system was created to address the needs of the broadcast video market.

In 1993, E&S helped Japanese arcade giant Namco with texture-mapping technology in Namco's System 22 arcade board that powered *Ridge Racer*. The help that E&S gave Namco was similar to the help that Martin Marietta gave Sega with the MODEL 2 board that powered Daytona USA and Desert Tank arcade games.

In 1998, E&S acquired AccelGraphics Inc, a manufacturer of computer graphics boards, for $52m. Since its launch in July 2002, the company's Digistar 3 system became the world's fastest selling Digital Theater system and is installed in upwards of 120 fulldome venues worldwide.

On May 9, 2006, E&S acquired Spitz Inc, a rival vendor in the planetarium market, giving the combined business the largest base of installed planetaria worldwide and adding in-house projection-dome manufacturing capability to E&S' offering.

In 2006, E&S sold its simulation business to Rockwell Collins. *(Wikipedia)*

Visual Display Systems

The early model-board display systems generally used TV screens in front of the replica cockpit to display the Out-The-Window (OTW) visual scene to the crew. Early computer-based image generator systems also used TV screens and sometimes projected displays. The focal distance of these displays was the distance of the screen from the crew, whereas objects in the real OTW visual scene were at a more distant focus, those close to the horizon being effectively at infinity.

Distant Focus Displays

In 1972, the Singer-Link company an outgrowth of the original Link Aviation, headquartered at Binghamton, New York, developed a display unit that produced an image at a distant focus. This took the image from a TV screen but displayed it through a collimating lens which had a curved mirror and a beamsplitter device. The focal distance seen by the user was set by the amount of vertical curvature of the mirror. These collimated display systems improved realism and depth perception for visual scenes that included distant objects.One of the challenges, however, was to obtain a realistic view from any crew member's eye point.

Optical infinity—Optical infinity was achieved by adjusting the focal distance so that it was above what is sometimes referred to as "optical infinity", which is generally taken as about 30 ft or nine meters. In this context, "optical infinity" is the distance at which, for the average adult person, the angle of view of an object at that distance is effectively the same for both the left and right eyes. For objects below this distance, the angle of view is different for each eye, allowing the brain to process scenes with a stereoscopic or three-dimensional result. The inference is that for scenes with objects which in the real world are at distances over about 9 meters / 30 feet, there is little advantage in using two-channel imagery and stereoscopic display systems in simulation display technology.

Collimated Monitor Design—The 1972 Singer-Link collimated monitors had a horizontal field of view (FoV) of about 28 degrees. In 1976, wider-angle units were introduced with a 35-degree horizontal FoV, and were called 'WAC windows', standing for 'Wide Angle Collimated', and this became a well-used term. Several "WAC Window" units would be installed in a simulator to provide an adequate field of view to the pilot for flight training. Single-pilot trainers would typically have three display units (center, left and right), giving a FoV of about 100 degrees horizontally and between 25 and 30 degrees vertically. One challenge is edge matching such that the picture area from all projectors is smooth and the edges are not perceptible.

Viewing Volume and user's Eye-point—For all of these collimated monitor units, the area from which the user had a correct view of the scene (the "viewing volume" from the user's "eye-point") was quite small. This was no problem in single-seat trainers because the monitors could be positioned in the correct position for the pilots' average eye-point. However, in multi-crew aircraft with pilots seated side-by-side, this led to each pilot only being able to see the correct outside-world scene through the collimated monitors that were positioned for that pilot's own eye-point. If a pilot looked across the cockpit towards the other crew member's display monitors, he/she saw distortions or even "black holes" because his/her viewing angle was outside the viewing volume established for the display units concerned. Personally, this situation would give me nausea. Clearly, for simulators with side-by side crew members, a system that gave correct cross-cockpit viewing was required.

Cross-cockpit Viewing

A breakthrough occurred in 1982 when the Rediffusion company of Crawley, UK, introduced their Wide-angle Infinity Display Equipment (WIDE) system. This used a curved mirror of large horizontal extent to allow distant-focus collimated viewing in a continuous, seamless, horizontal display for pilots seated side-by-side. (Figure 73). The Out-The-Window (OTW) image was back-projected on a screen above the replica cockpit, and it was the reflection from this screen that was viewed by the pilots. To avoid the weight and fragility of using a large glass mirror, the reflective material appeared on a flexible mylar sheet. When the simulator is in operation, an accurate shape for the flexible sheet is maintained by its attachment to a shaped former by suction pressure produced by a small vacuum pump. The other major flight simulation companies now produce their own types of mirror-based cross-cockpit displays and these are now utilized in most full flight simulators of Regulatory Levels C and D. The original cross-cockpit display systems used three projectors mounted on top of the replica cockpit and had anFoVof 150 degrees horizontally by 30 degrees vertically. With five projectors the horizontal FoV could be extended to 220 degrees. Developments have allowed these figures for three- and five-projector systems to be extended to 180 degrees with three projectors and 240 degrees with five. *(Wikipedia)*

Figure 73—Cross-cockpit Collimated Displays

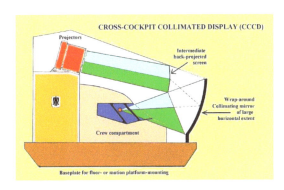

Cross-Cockpit Collimated display system - flight simulator application

Visual systems continue to improve in fidelity and all dimensions (Figure 74). Notice the high resolution display with sufficient detail to perform air-to-air combat maneuvers, formation flying or carrier landings. This is the type of realisim pilots are looking for when they submit to simulator substitution for some aircraft time.

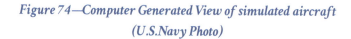

Figure 74—Computer Generated View of simulated aircraft
(U.S.Navy Photo)

Computers

In simulators, we began with the Link pneumatic system, moved to analog computers and then on to digital computers to solve the equations of motion and perform other computational functions. A research project led by the University of Pennsylvania under a contract to the Naval Training Device Center (NTDC) in Port Washington, New York, called UDOFT (Universal Digital Operational Flight Trainer) resulted in the first digital computer targeted for flight simulators. That was in 1950 when digital computers were inadequate for high speed computing and the input/output devices were too awkward to be applied in real time simulation. UDOFT resulted in a special purpose computer that was actually manufactured by Sylvania. Meanwhile, recognizing the eventual power of digital computers, Link designed their own Mark I computer for simulators. By the 1970's, all new trainers for simulators incorporated digital computers. By that time there were several companies competing for the same business with their general purpose mainframe computers. I recall working on aviation trainers to support anti-submarine warfare training that utilized Packard-Bell delay line computers. Today, simulators do not rely on a single computer but a network of very powerful and smaller computers. An interesting side note: when digital computers were introduced to simulators, the computer people made a point of not referring to the "memory" in a computer. They preferred to call it "storage" instead. They were so scared that the general public would begin to humanize computers to the point where they would be afraid of them.

Motion Systems

The argument goes on about motion systems concerning the value of simulator motion to training effectiveness. The airlines insist on simulator motion for a certain level of training. Ed Link was a strong proponent of using motion in flight simulators. He professed that motion was necessary to give the student pilot the true sensations of flight. Motion systems have an interesting and controversial history.

The motion system in the 1929 Link Trainer design gave movements in pitch, roll and yaw, but the payload (weight of the replica cockpit) was limited. For flight simulators with heavier cockpits, the Link Division of General Precision Inc. (later part of Singer Corporation and now part of L-3 Communications) in 1954 developed a system where the cockpit was housed within a metal framework that provided three degrees of displacement in pitch, roll, and yaw. It was found that six jacks in the appropriate layout could produce all six degrees of freedom that are possible for a body that can freely move. These are the three angular rotations pitch, roll and yaw, and the three linear movements heave (up and down), sway (side to side) and surge (fore and aft). The design of such a 6-jack (hexapod) platform was first used by Eric Gough in 1954 in the automotive industry and further refined by Stewart in a 1966 paper to the UK Institution of Mechanical Engineers and named the Stewart platform.

From about 1977, aircraft simulators for Commercial Air Transport (CAT) aircraft were designed with ancillaries such as Instructor Operating Stations (IOS), computers, etc., being placed on the motion platform along with the replica cockpit, rather than being located off the motion platform. (*Wikipedia-Flight Simulators*)

Other lower cost approaches to motion and sensory cueing have emerged with dynamic motion seats. There are a few of them such as that manufactured by Acme Worldwide Enterprises in Albuquerque NM and Industrial Smoke and Mirrors in Orlando, FL. These seats have proven to be effective in studies conducted by the Manned Flight Center in Patuxant River, MD and NAWCTSD in Orlando. At one time, after retiring I worked with Acme as their Orlando Representative.

Simulation vs. Stimulation

As part of the systems engineering process for simulators, one of the decisions to be made is should we "simulate" aircraft systems ot should we "stimulate" the aircraft system? Simulator designers have formalized processes to look at all of the factors that come into play in making a decision to simulate or stimulate. There are many factors to be considered in that decision process such as: are the aircraft systems available, can you guarantee delivery of aircraft parts on time, cost, etc. The same holds true for shipboard or undersea platforms.

Embedded Training

At times, a requirement can exist for putting training modes in the operational system. This has been accomplished for tracked and other vehicles, ships, submarines and aircraft. For aircraft, there is a safety issue in incorporating training modes in the pilot's position in operational aircraft. Suppose a pilot thought he or she was in a training mode but was mistaken, firing a missile at a real target. For ship or submarine or ground operational equipment, embedding training in the prime system can be very cost effective. Trainees will already be on board and require no separate facilities and movement of the ship to a training range is not necessary. Pierside Trainers. (Figures75 and 76) is the step between institutional training at a schoolhouse and training that can be given underway. With Pierside Trainers, there is an electrical connection between the trailers that house the "stimulation" hardware and software located on the "pier" with the on-board equipment that provides the interface to the actual shipboard system.

Figure 75—DEVICE 20B4—Pierside Mobile Combat System Team Trainers

The Device 20B4's were fielded in the early 1980s to support afloat combat systems. The device consists of two elements:

1) A mobile van housing a computational system, data/voice communication equipment, problem control and monitor equipment and

2) A carry on compliment of interface equipment to provide data/signal sampling and interface with onboard sensors.

The tactical environment including air, surface, and subsurface threat conditions that are presented to the combat systems teams through stimulation of onboard sensor systems, and by providing external tactical data. All onboard radar, sonar and weapon launch equipment are stimulated with threat target data. The weapon trajectory and resultant damage assessment are modeled and appropriate status signals generated to drive operator display equipment. *(NAWCTSD website)*

Figure 76—Pierside Trainer and FFG-7 Ship

Pierside trainers are but one example of training systems developed to support our surface and submarine community. Additional surface and submarine trainers are shown in Figures 77 and 78.

Figure 77—Device14A12(J) Team Tactics Trainer (US Navy)

Figure 78—Trident Ship Control Trainer (US Navy)

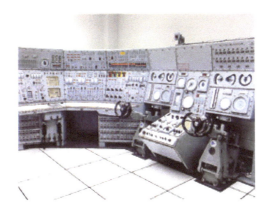

Physiology Training

At times it is necessary to subject military trainees to conditions and stresses they might experience in the real world. A strictly "virtual world" just isn't enough to give the trainee a feel for what they might be subjected to at their duty station. For example, trainees in basic training must go through the *gas chamber,* learning how to put on the gas mask, clear existing gas and breathe effectively. Tear gas or other harmless gasses are used in the training. There are training devices that simulate forces on the body by producing those forces of the magnitude and direction found in the real world. The centrifuge represents this category. Realistic conditions that might occur on board ship can also be simulated. Trainers are used for damage control training by simulating torpedo, missile or other weapon damage to a ship or submarine, with sea water gushing into the compartment-requiring quick, focused and even team attention. Some examples follow.

Bailout Trainer

When I first came to work with the Navy, I was told that there was a trainer used to train pilots how to parachute from an SNJ Aircraft. (Figure 79) The improvised "trainer" was built by some creative sailors who felt they could satisfy a real training need. The problem they set out to solve was that pilots egressing from a seriously damaged airplane were striking the tail section of the aircraft. The trainer gave them practice for catapaulting out far enough from the cockpit to miss colliding with the aircraft. Again, necessity breeds invention-no matter how simple it may be.

Figure 79—SNJ Aircraft Bailout Trainer (Dick Adkins)

Barany Chair (Disorientation)

The Barany chair or Bárány chair, named for the Hungarian physiologist Bárány Róbert, is a device used for aerospace physiology training, particularly for student pilots. (Figure 80) The subject is placed in the chair, blindfolded, and then spun about the vertical axis while keeping his or her head upright or tilted forward or to the side. The subject is then asked to perform tasks such as determine his or her direction of rotation while blindfolded, or rapidly change the orientation of the head, or attempt to point at a stationary object without blindfold after the chair is stopped. The chair is used to demonstrate spatial disorientation effects, proving that the vestibular system is not to be trusted in flight. Pilots are taught that they should instead rely on their flight instruments.

Bárány used this device in his ground-breaking research into the role of the inner ear in the sense of balance, which won him the 1914 Nobel prize. *(Wikipedia).*

Figure 80—Barany Chair in use (Laughlin AFB Photo)

Human Centrifuge Training

There are several centrifuges used by the United States military and NASA to create actual g forces on the body. Russia has an interesting centrifuge used to train cosmonauts. (Figure 81)

Figure 81—TsF-18 centrifuge at the Yuri Gagarin Cosmonauts Training Center

Human centrifuges are exceptionally large systems that test the reactions and tolerance of pilots and astronauts to acceleration above those experienced in the Earth's gravity.

The US Air Force at Brooks City-Base in San Antonio, Texas, utilizes a human centrifuge. The centrifuge at Brooks is operated by the aerospace physiology department for the purpose of training and evaluating fighter pilots and Weapon Systems Officers for high-*g* flight in Air Force fighter aircraft.

The use of large centrifuges to simulate a feeling of gravity has been proposed for future long-duration space missions. Exposure to this simulated gravity would prevent or reduce the bone decalcification and muscle atrophy that affect individuals exposed to long periods of freefall. An example of this can be seen aboard the *Discovery* spacecraft in the film *2001: A Space Odyssey*.

Parachute Disentanglement Trainer

The US Navy has a family of physiology training devices. One of the most challenging is the Parachute Disentanglement Trainer. This device esentially drags the trainee through a swimming pool, simulating the wind blowing the parachute after bail-out and a water landing, and challenges the pilot to deal with that situation. (Figure 82)

Figure 82—"Archie Arcidiacono" NTDC Employee, in the Parachute Disentanglement Trainer (U.S. Navy)

Dilbert Dunker

The Dilbert Dunker is a device for training pilots on how to correctly escape from a submerged plane that has crashed or landed on the ocean or other body of water. It was invented by Wilfred Kaneb, a nautical engineer, during World War II. The word Dilbert is not listed in the dictionary although the term is used in the United States Navy to define a person who is slow witted and incapable of getting things done correctly. It is combined with "dunker" because a mockup cockpit (SNJ Aircraft in the 40's and 50's) is sent down a 45 degree rail from a high stand at the deep end of the training pool, and at the end of the run under water it flips inverted to simulate a water ditching.(Figure 83). The preflight student must detach the communication wire from the helmet, release the seat and shoulder harness, dive still deeper and swim away from the "aircraft" at a 45 degree angle to the surface for the purpose of assuming that the water around an actual situation has burning fuel on the waters surface.You may rememeber the device was featured in the 1982 film, *An Officer and a Gentleman*, starring Richard Gere and Debra Winger.

Figure 83—Dilbert Dunker (Circa 1950) (Wikipedia)

High Altitude Low Pressure Trainer

The High Altitude Low Pressure Trainer has been around since the 1930's with the purpose of having pilots and aircrews become familiar with what happens with oxygen deprivation. What they do is put the trainees in an air-tight chamber and essentially evacuate the air in the chamber. Various techniques are used such as bringing the simulated altitude up to about 18,000 feet and have the trainees remove their masks and let them experience what happens when their oxygen level is reduced. They give air crews tasks to do and let them experience the "numbness" of the brain that occurs in that deprived condition. At one time, the instructors would initiate a rapid decompression of the chamber which has major effects on most people. I understand that this is no longer done for safety reasons. The chambers were also used to filter-out the trainees who could not adapt to that rarified environment.

During World War II it was difficult to convince aircrews that they needed to keep their oxygen masks on at high altitude. The masks were uncomfort-able but necessary. One demonstration reported during that era was to put the gunner and pilot in the chamber. The gunner had one of those amuse-ment machine guns-the kind we found in corner drug stores, arcades and cigarette shops that were coin-operated. The game objective was to shoot down as many moving airplanes possible with the gun that fired a light beam. Gunners would do great with oxygen flowing to their brains. Take away the oxygen at a simulated 15,000 feet and the gunner did terribly. Both the gunner and pilot learned a very important lesson about the brain's per-formance without oxygen. It doesn't work as well!

Gaming

"Gaming" to me implies interaction between two or more players with a common objective of "winning"-however "winning" might be defined. The adversary need not be a live person but could be computer generated. Actual games go back to early forms of chess and the use of sand tables. However, I will not dwell on the physical, I intend to focus on video games.

The first evidence of a video game goes back to 1947 when a patent for a "cathode tube amusement device" was filed by Thomas T. Goldsmith and Estle Ray Mann. The objective of the game was to simulate firing at an airborne target. I recall a "Bullpup" trainer that was developed for NTDC in the 1960's that had the same objective. In 1949–1950, Charley Adama created a "Bouncing Ball" program for Massachusetts Institute of Technology's (MIT) Whirlwind computer. While the program was not yet interactive, it was a precursor to games soon to come.

In 1951,while developing television technologies for New York based electronics company Loral, inventor Ralph Baer came up with the idea of using the lights and patterns he used in his work as more than just calibration equipment. Ralph Baer has been named by some as the "father of video games". (Figure 84). He realized that by giving an audience the ability to manipulate what was projected on their television sets, their role changed from passive observation to interactive manipulation. When he took this idea to his supervisor, it was quickly squashed. This was not uncommon in those days when employees were supposed to focus only on what the job called for. Innovation or thinking outside the box was considered a waste of time. A graphical version of tic-tac-toe, was created by A.S. Douglas in 1952 at the University of Cambridge, in order to demonstrate his thesis on human-computer interaction. It was developed on the EDSAC computer, which uses a cathode ray tube as a visual display to display memory contents. (Memory was not "politically correct" at the time because it humanized a machine.) In Douglas' game, the player competes against the computer.

Figure 84—Ralph Baer—Father of Video Games (source unknown)

In 1958, William Higinbotham created a game using an oscilloscope and analog computer. Titled *Tennis for Two*, it was used to entertain visitors of the Brookhaven National Laboratory in New York. Unfortunately, oscilloscopes were use primarily in the laboratory or for maintenance and not found in the home. *Tennis for Two* showed a simplified tennis court from the side, featuring a gravity-controlled ball that needed to be played over the "net," unlike its successor—*Pong*. The game was played with two box-shaped controllers, both equipped with a knob for trajectory and a button for hitting the ball. *Tennis for Two* was exhibited for two seasons before it was dismantled in 1959. Today *Pong* is considered an "old school computer game" played on cellphone and tablets.

The majority of early computer games ran on universities' mainframe computers in the United States and were developed by individuals as a hobby. The limited accessibility of early computer hardware meant that these games were small in number and forgotten by posterity.

In 1959–1961, a collection of interactive graphical programs were created on the TX-0 machine at MIT as follows:

- *Mouse in the Maze*: allowed players to place maze walls, bits of cheese, and, in some versions, martinis using a light pen. One could then release the mouse and watch it traverse the maze to find the goodies.

- *HAX*: By adjusting two switches on the console, various graphical displays and sounds could be made.
- *Tic-Tac-Toe*: Using the light pen, the user could play a simple game of tic-tac-toe against the computer.

In 1961, a group of students at MIT, including Steve Russell, programmed a game titled *Spacewar!* on the PDP-1, a new computer at the time. The game pitted two human players against each other, each controlling a spacecraft capable of firing missiles, while a star in the center of the screen created a large hazard for the crafts. The game was eventually distributed with new Digital Equipment Corporation (DEC) computers and traded throughout the then-primitive Internet. *Spacewar!* is credited as the first influential computer game.

Also in 1961, John Burgeson wrote the first computer baseball simulation game on an IBM 1620 Computer at the IBM facility in Akron, Ohio. Users picked a lineup and could then watch the results of their simulated game printed out by the computer. This was a unique approach for that time and set the stage for user involvement in selecting teams or players in a simulation.

In the early 1960s, I came up with a game based on the paper game "Battleship". It was rather unique because of the simple logic employed and the novel packaging concept. (Figure 85). I will not dwell on the fact that my friend gave it to two toy companies who expressed no interest in it. However, it came out in the market a year or so later. That was my first lesson in protecting intellectual property!

Figure 85—Electronic Game Based on "Battleship"

In 1966, Ralph Baer engaged co-worker Bill Harrison in the project, where they both worked at military electronics contractor Sanders Associates in Nashua, New Hampshire.

I had occasion to visit Sanders when they were under contract to the Navy in Orlando to deliver a simulator. At that time, they must have made a decision to diversify and seek non-military business. They created a simple video game named *Chase*, the first to display on a standard television set. With the assistance of Baer, Bill Harrison created the light gun. Baer and Harrison were joined by Bill Rusch in 1967, an MIT graduate who was subsequently awarded several patents for the TV gaming apparatus. Development continued, and in 1968 a prototype was completed that could run several different games such as table tennis and target shooting. After months of secretive laboring between official projects, the team was able to bring an example with true promise to Sanders' R & D department. By 1969, Sanders was showing off the world's first home video game console to manufacturers.

In 1969, AT&T computer programmer Ken Thompson wrote a video game called *Space Travel* for the Multics operating system. This game simulated various bodies of the solar system and their movements and the player could attempt to land a spacecraft on them. AT&T pulled out of the Multics project, and Thompson ported the game to FORTRAN code running on the

GECOS operating system of the General Electric GE 635 mainframe computer. Runs on this system cost about $75 per hour, and Thompson looked for a smaller, less expensive computer to use. He found an underused PDP-7, and he and Dennis Ritchie started porting the game to PDP-7 assembly language. In the process of learning to develop software for the machine, the development process of the Unix operating system began, and *Space Travel* has been called the first UNIX application.

The next migration of video games was to the **arcades**. In 1971, *Galaxy Game* was installed at Stanford University. Based on *Spacewar!*, this was the first coin-operated video game. Only one was built, using a DEC PDP-11 computer and vector display terminals. In 1972, it was expanded to be able to handle four to eight consoles.

Also in 1971, Nolan Bushnell and Ted Dabney created a coin-operated arcade version of *Spacewar!* and called it *Computer Space*. Nutting Associates bought the game and manufactured 1,500 *Computer Space* machines, with the release taking place in November 1971. The game was unsuccessful due to its steep learning curve. Later it was a landmark game as the first mass-produced video game and the first offered for commercial sale.

Bushnell and Dabney founded Atari, Inc. in 1972, before releasing their next game: *Pong*. *Pong* was the first arcade video game with widespread success. This was a very popular game for young and not-so-young alike. The game is loosely based on table tennis: a ball is "served" from the center of the court and as the ball moves towards their side of the court each player must maneuver their paddle to hit the ball back to their opponent. Atari sold over 19,000 *Pong* machines, creating many imitators.

Another significant game was *Gun Fight*, an on-foot, multi-directional shooter, designed by Tomohiro Nishikado and released by Taito in 1975. It depicted game characters, game violence, and human-to-human combat, controlled using dual-stick controls. The original Japanese version was based on discrete logic, which Dave Nutting adapted for Midway's American release using the Intel 8080 computer, making it the first video game to use a microprocessor. This later inspired original creator Nishikado to use a microprocessor for his 1978 blockbuster hit, *Space Invaders*.

Now, it was time to move into *home consoles*. We are now on the **seventh generation** of home consoles that are affordable on the commercial market.

The first home "console" system was developed by Ralph Baer and his associates. Development began in 1966 and a working prototype was completed by 1968 (called the "Brown Box") for demonstration to various potential licensees, including GE, Sylvania, RCA, Philco, and Sears, with Magnavox eventually licensing the technology to produce the world's first home video game console. The system was released in the USA in 1972 by Magnavox, called the Magnavox Odyssey. (Figure 86).

Figure 86—The Magnavox Odyssey (Wikipedia)

The Odyssey used cartridges that mainly consisted of jumpers that enabled/disabled various switches inside the unit, altering the circuit logic (as opposed to later video game systems that used programmable cartridges). This provided the ability to play several different games using the same system, along with plastic sheet overlays taped to the television that added color, play-fields, and various graphics to 'interact' with using the electronic images generated by the system. A major marketing push, featuring TV advertisements starring Frank Sinatra, helped Magnavox sell about 100,000 Odyssey consoles that first year.

Philips bought Magnavox and released a different game in Europe using the Odyssey brand in 1974 and an evolved game that Magnavox had been developing for the US market. Over its production span, the Odyssey system achieved sales of 2 million units.

The generation opened early for handheld consoles, as Nintendo introduced their Nintendo DS and Sony premiered the PlayStation Portable (PSP) within a month of each other in 2004. While the PSP boasted superior graphics and power, following a trend established since the mid-1980s, Nintendo gambled on a lower-power design but featuring a novel control interface. The DS's two screens, one of which was touch-sensitive, proved extremely popular with consumers, especially young children and middle-aged gamers, who were drawn to the device by Nintendo's *Nintendogs* and *Brain Age* series, respectively. While the PSP attracted a significant portion of veteran gamers, the DS allowed Nintendo to continue its dominance in handheld gaming. Nintendo updated their line with the Nintendo DS Lite in 2006, the Nintendo DSi in 2008 (Japan) and 2009 (Americas and Europe), and the Nintendo DSi XL while Sony updated the PSP in 2007 and again with the smaller PSP Go in 2009. Nokia withdrew their N-Gage platform in 2005 but reintroduced the brand as a game-oriented service for high-end smartphones on April 3, 2008.

In console gaming, Microsoft stepped forward first in November 2005 with the Xbox 360, and Sony followed in 2006 with the PlayStation 3, released in Europe in March 2007. Setting the technology standard for the generation, both featured high-definition (HD) graphics over HDMI connections, large hard disk-based secondary storage for save games and downloaded content, integrated networking, and a companion on-line gameplay and sales platform, with Xbox Live and the PlayStation Network, respectively. Both were formidable systems that were the first to challenge personal computers in power (at launch) while offering a relatively modest price compared to them. While both were more expensive than most past consoles, the Xbox 360 enjoyed a substantial price edge, selling for either $300 or $400 depending on model, while the PS3 launched with models priced at $500 and $600. Coming with Blu-ray Disc and Wi-Fi, the PlayStation 3 was the most expensive game console on the market since Panasonic's version of the 3DO, which retailed for little under $700.

Nintendo would release their Wii console shortly after the PlayStation 3's launch, and the platform would put Nintendo back on track in the console race. While the Wii had lower technical specifications than both the Xbox 360 and PlayStation 3, only a modest improvement over the GameCube and the only 7th-gen console not to offer HD graphics, its new motion control was much touted, and its lower pricepoint of around $200–$250 appealed to budget-conscious households. Many gamers, publishers, and analysts initially dismissed the Wii as an underpowered curiosity, but were surprised as the console sold out through the 2006 Christmas season, and remained so through the next 18 months, becoming the fastest selling game console in most of the world's gaming markets.

The Wii's major strength was its appeal to audiences beyond the base of "hardcore gamers", with its novel motion-sensing, pointer-based controller design allowing players to use the Wii Remote as if it were a golf club, tennis racket, baseball, sword, or steering wheel. The intuitive pointer-based navigation was reminiscent of the familiar PC mouse control, making the console easier for new players to use than a conventional gamepad. The console launched with a variety of first-party titles meant to showcase what the system could do, including the commonly bundled *Wii Sports*, the party game *Wii Play*, and *The Legend of Zelda: Twilight Princess* which received widespread critical acclaim for its art, story and gameplay, including its intuitive sword-and-shield use of the motion-sensing controllers. Third-party support, on the other hand, was slower in coming; game designers, used to the conventional gamepad control sets, had difficulty adapting popular franchises to the heavily motion-based Wii controller, and some didn't bother, preferring instead to concentrate on what they felt to be their core audience on the Xbox 360 and PlayStation 3. Of the top 15 best-selling games for Wii, 12 were developed by Nintendo's in-house EAD groups.

In June 2009, Sony announced that it would release its PSP Go for $249.99USD on October 1 in Europe and North America, and Japan on November 1. The PSP Go was a newer, slimmer version of the PSP, which had the control pad slide from the base, where its screen covers most of the front side. *(Wikipedia)*

Serious Games

Simulation technology, as I have brought out in previous chapters, experienced a number of "spin-off's" from military and space applications to a a myriad of other uses. Essentially, the Department of Defense (DOD) paid the bills for developing the technology and other communities would take the basic technology and modify or enhance it for the other application. Just the reverse is taking place with gaming technology. For example, Microsoft Flight Simulator was developed and sold as a game for users of all ages. I play it as do my grandchildren. You can actually learn to fly an aircraft using this program. It is so effective that a student in Naval Undergraduate Pilot Training in Pensacola actually improved his grades and progress by flying Microsoft Flight Simulator after class in the barracks. It is now an integral part of flight training.

The military has adopted other "serious games" into their curriculum. This constitutes a "spin-on" of technology. In this case, commercial technology is adopted by the military with the commercial buyer actually paying for the cost of development.

Serious games is not a new idea. Military officers have been using war games in order to train strategic skills for a long time. One early example of a serious game is a 19th century Prussian military training game called *Kriegsspiel*, the German name for wargame. I recall working with the Marines years ago when board games were in demand by the Corps. These very simple games provided a forum for studying and developing battlefield tactics, demonstrating that the technology need not be overly complex to be of value. To-day, one of the most popular serious games, particularly with the Army and Marine Corps, is Virtual Battlespace 2.(VBS2). (Figure 87)

Figure 87—Bohemia Interactive's VBS2 (Bohemia Website)

VBS2 offers realistic battlefield simulations and the ability to operate land, sea, and air vehicles. Instructors may create new scenarios and then engage the simulation from multiple viewpoints. The squad-management system enables partici-pants to issue orders to squad members.

VBS2 may be used to teach doctrine, tactics, techniques, and procedures during squad and platoon offensive, de-fensive, and patrolling operations. VBS2 delivers a synthetic environment for the practical exercise of the leadership and organizational behavior skills required to successfully execute unit missions.

VBS2 is suitable for training small teams in urban tactics, entire combat teams in combined arms operations or even non-mil-itary usage such as emergency response

The simulation engine driving VBS2 is Real Virtuality 2, devel-oped by Bohemia Interactive. VBS2 allows a user to develop large terrain areas, over 10,000 square kilometres (3,900 sq mi)

in size (at any terrain resolution) and populate the terrain area with millions of objects in accordance with VMAP shape data, and then texture-map the entire representation with high-resolution satellite imagery or aerial photography.

Once the terrain representation is exported into VBS2, the simulation engine will provide a simulation of the real world, incorporating moving trees and grass, ground clutter, ambient animal life, shadows, dynamic lighting, weather and time of day. Weapon platforms are capable of thermal imaging, simulation of fire control systems and turret override. Multiple vehicle turrets are possible. Weapon ballistics have been improved.

The After-Action-Review module allows detailed review of a completed training mission, with every player, AI, vehicle movement being recorded, as well as any bullet path and any destruction to objects or terrain.

The VBS1 High Level Architecture/Distributed Interactive Simulation (HLA/DIS) gateway is updated and improved for VBS2 to meet HLA and DIS compliance. *(Wikipedia)*

Networking

Up until the 1980s, the focus had been on individual and own-ship training. In other words, we became quite good at training pilots and aircrew on board a particulat tank, airplane, submarine or ship. The next logical step was to train teams who might be in separate simulators and even geographically separated.

Jack Thorpe of the Defense Advanced Research Projects Agency (DARPA) saw the need for networked multi-user simulation. Interactive simulation equipment was very expensive, and reproducing training facilities was likewise expensive and time consuming. In the early 1980s, DARPA decided to create a prototype research system to investigate the feasibility of creating a real-time distributed simulator for combat simulation. Simulator Networking (SIMNET), the resulting application, was to prove both the feasibility and effectiveness of such a project *(Pimental and Blau 1994).*

Training using actual equipment was extremely expensive and danger-ous. Being able to simulate certain combat scenarios, and to have partici-pants remotely located rather than all in one place, hugely reduced the cost of training and the risk of personal injury *(Rheingold 1992).* The concept of performance-based training was pushed very strongly by another visionary in our field. General Paul Gorman, Training and Doctrine Command of the Army, essentially revolutionized training within the Army. He appreci-ated the value of simulation and became its strongest advocate. Thus, when SIMNET came along, he understood the potential value of the networked simulation and became a prime mover in distributed training for the Army. He worked closely with PM TRADE's Bill Marrioletti and Ernie Smart (one of his former colonels) at UCF's Institute for Simulation and Training in getting simulation into various training programs..

Long-haul networking for SIMNET was run originally across multiple 56 kbit/s dial-up modem lines, using parallel processors to compress packets over the data links. This traffic contained not only the vehicle data but also compressed voice.

Since this was a networked simulation, each simulation station needed its own display of the shared virtual environment. The display stations them-selves were mock-ups of certain tank and aircraft control simulators, and they were configured to simulate actual conditions within the combat vehi-cle. The tank simulators, for example, could accommodate a full four-person crew complement to enhance the effectiveness of the training. The network was designed to support up to several hundred users at once. The fidelity of the simulation was such that it could be used to train for mission scenarios and tactical rehearsals for operations performed during the U.S. actions in Desert Storm in 1992 *(Robinett 1994).*

SIMNET used the concept of "dead reckoning" to correlate the positions of the objects and actors within the simulated environment. In other words, "dead reckoning" would extrapolate a curve to predict the next position. Of course, this would introduce errors especially for fast-moving platforms such as a jet aircraft. For this initial proof-of-concept, however, it was satisfactory. Duncan (Duke) Miller, the BBN SIMNET program manager, first used this term, which goes back to the earliest days of ship navigation, to explain how simulators were able to communicate state change information to each other while minimizing network traffic. Essentially, the approach involves calculating the current position of an object from its previous position and velocity (which is composed of vector and speed elements) *(Pimental and Blau 1994)*. The SIMNET protocols provided that whenever the true state of a simulator deviated by more than a certain threshold from its state as computed by dead reckoning, the simulator was obligated to send out a new state update message.

SIMNET protocols and SIMNET-based training systems use in the First Gulf War demonstrated the success of the networked simulation, and its legacy was viewed as proof that realtime interactive networked cooperative virtual simulation is possible for a large user population. Later, the Terrestrial Wideband Network (a high speed descendant of the ARPANET that had computers running at T1 speeds) was used to carry traffic. This network remained under DARPA after the rest of ARPANET was merged with NSFNet and the ARPANET was decommissioned *(Rheingold)*

The follow-on protocols to SIMNET were called Distributed Interactive Simulation (DIS); the primary U.S. Army follow-on program was the Close Combat Tactical Trainer (CCTT).

The SIMNET-D (Developmental) program used simulation systems developed in the SIMNET program to perform experiments in weapon systems, concepts, and tactics. It became the Advanced Simulation Technology Demonstration (ASTD) program. It fostered the creation of the Battle Labs across the US Army, including the Mounted Warfare TestBed at Ft Knox, Ky, the Soldier Battle Lab at Ft Benning, GA, the Air Maneuver Battle Lab at Ft Rucker, AL, the Fires Battle Lab at Ft Sill, OK. *(Wikipedia)*.

SIMNET became the first successful implementation of a large-scale, real-time, man-in-the-loop simulator networking for team training and mission rehearsal in military operations. As is the case of the introduction of any new technology, there needs to be a strong advocate who is in a policy-making position. Along comes Gen Paul Gorman who was the prime mover or the Army in getting distributed simulation off the ground. At the University of Central Florida's Institute for Simulation and Training, Ernie Smart, a retired Army Colonel, demonstrated the SIMNETs to hundreds of visitors and was instrumental in introducing the new technology. (Figure 88).

Figure 88—Ernie Smart Demonstrating SIMNET at IST

(Intellectual property of University of Central Florida Institute for Simulation and Training (IST))

The earliest successes that came through the SIMNET program was the demonstration that geographically dispersed simulation systems could support distributed training by interacting with each other across network connections. That represented a paradigm shift for our technology. We began to think in terms of total battlefield assets rather than single platform capabilities. I am sure a few visionaries were already thinking L-V-C *(Live-Virtual-Constructive)*.

The Aggregate Level Simulation Protocol (ALSP) extended the benefits of distributed simulation to the force-level training community so that different aggregate-level simulations could cooperate to provide theater-level experiences for battle-staff training. The ALSP has supported an evolving "confederation of models" since 1992, consisting of a collection of infrastructure software and protocols for both inter-model communication through a common interface and time advance using a conservative Chandy-Misra based algorithm.

At about the same time, the SIMNET protocol evolved and matured into the Distributed Interactive Simulation (DIS) Standard. DIS allowed an increased number of simulation types to interact in distributed events, but was primarily focused on the platform-level training community. DIS provided an open network protocol standard for linking real-time platform-level wargaming simulation.

In the mid 1990s, the Defense Modeling and Simulation Office (DMSO) sponsored the High Level Architecture (HLA) initiative. Designed to support and supplant both DIS and ALSP, investigation efforts were started to prototype an infrastructure capable of supporting these two disparate applications. The intent was to combine the best features of DIS and ALSP into a single architecture that could also support uses in the analysis and acquisition communities while continuing to support training applications.

The DOD test community started development of alternate architectures based on their perception that HLA yielded unacceptable performance and included reliability limitations. The real-time test range community started development of the Test and Training Enabling Architecture (TENA) to provide low-latency, high-performance service in the hard-real-time application of integrating live assets in the test-range setting. TENA, through its common infrastructure, including the TENA Middleware and other complementary architecture components, such as the TENA Repository, Logical Range Archive, and other TENA utilities and tools, provides the architecture and software implementation and capabilities necessary to quickly and economically enable interoperability among range systems, facilities, and simulations.

Similarly, the U.S. Army started the development of the Common Training Instrumentation Architecture (CTIA) to link a large number of live assets requiring a relatively narrowly bounded set of data for purposes of providing After Action Reviews (AARs) on Army training ranges in the support of large-scale exercises.

Other efforts that make the Live, Virtual Constructive (LVC) architecture space more complex include universal interoperability software packages such as OSAMS (or CONDOR developed and distributed by commercial vendors.

As of 2010 all of the DoD architectures remain in service with the exception of SIMNET. Of the remaining architectures: CTIA, DIS, HLA, ALSP and TENA, some are in early and growing use (e.g., CTIA, TENA) while others have seen a user-base reduction (e.g., ALSP). Each of the architectures is providing an acceptable level of capability within the areas where they have been adopted. However, DIS, HLA, TENA, and CTIA-based federations are not inherently interoperable with each other. When simulations rely on different architectures, additional steps must be taken to ensure effective communication between all applications. These additional steps, typically involving interposing gateways or bridges between the various architectures, may introduce increased risk, complexity, cost, level of effort, and preparation time. Additional problems extend beyond the implementation of individual simulation events. As a single example, the ability to reuse supporting models, personnel (expertise), and applications across the different protocols is limited. The limited inherent interoperability between the different protocols introduces a significant and unnecessary barrier to the integration of live, virtual, and constructive simulations (*Wikipedia—Live, Virtual, Constructive*).

Today, it is possible to link together simulators in a manner equal to the way the platforms operate in their missions through Distributed Mission Training (DMT). In other words, "we train as we fight and fight as we train." An excellent example is the DMT for the F-16 aircraft. In operations, the F-16s conduct missions in a four ship configuration. So, it is logical that the pilots train in that same configuration. Through DMT, it is now possible to train as individual pilots but as a part of a tactical team. (Figure 89).

A great deal of credit for developing the standards for networking goes to UCF/IST, the Simulation Interoperability Standards Organization (SISO) and the Institute of Electrical and Electronic Engineers (IEEE). These organizations had the vision to establish the necessary standards, infrastructure and procedures to enable training assets to be networked.

Figure 89—Four F-16 Trainers networked (U.S. Air Force Photo)

Analysis
Analyze Airport Operations
One of the distinct advantages of simulation is that it enables us to analyze complex problems gaining unusual insight into the system being studied. It has been said that simulation is the new calculus. Further, when simulation is coupled with visualization, an entirely new perspective is achieved. It must be qualified though that the simulation is only as good as the math-

ematical model we start with. An example is to study airport operations using simulation. (Figure 90)

Airport terminals have dramatically changed after September 11th, primarily due to the tightened security measures. These changes had a major impact on passenger arrival patterns, passenger flows, space allocation, processing times, and waiting times. In turn, it impacted terminals' performance, levels of service, and the overall passenger experience. Airport planners and decision makers required a decision support tool that can quickly evaluate the impact of the often changing security regulations and the decisions to counterpart these changes on the airport's level of service. Discrete event simulation is an important system analysis technique that can assist airport planners and decision makers in improving the airport and passenger flow.

Figure 90—Passenger Flow through Orlando International Airport
(Productivity Apex)

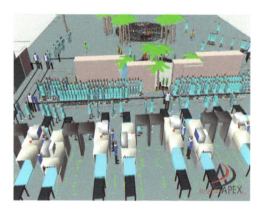

Oil Spill Analysis

Dr.Poojitha Yapa of Clarkson University developed mathematical models that describe the movement of fluids (oil) in the water. His models brought together the factors influencing oil movement such as wind, tide, viscosity, etc. into a simulation of an oil spill. This simulation is helpful in determining where first responders, for example, might position themselves and cleaning equipment to minimize the damage caused by an oil spill. His models were validated during the Gulf of Mexico disaster. (Figure 91). Models and simulations like this can be predictive or be used for mission rehearsal.

Figure 91—Oil Disaster Simulation based on Results of CDOG/ADMS Model

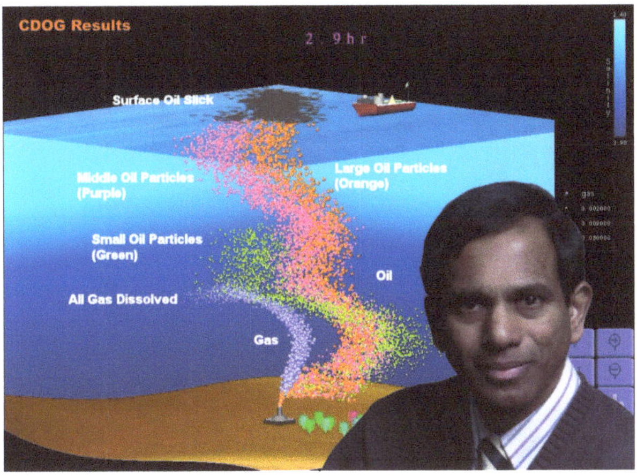

(The copyrights for the image remain with Dr. Poojitha Yapa, Clarkson University)

Terrorist Attack Simulation

Dr. Christoph M. Hoffman at Purdue University developed an interesting model and simulation (Figures 92 and 93) of a terrorist attack on a building similar to the World Trade Center in New York City. His model brings together the building design and Finite Element Analysis (FEA) techniques to study the effects of an airplane crashing into a building. The output visualization is not simply an animation of the event. The output is an engineering result that gives great insight into the building design and how it might be changed to minimize damage and loss of life, plus improved design for more efficient escape is possible. A few years ago, Dr. Hoffman attended the Interservice/Industry Training, Simulation and Education Conference in Orlando and briefed about 35 science teachers of his work. The teachers had no idea such an approach was even possible. Perhaps more importantly, those teachers return to the classroom where they can demonstrate the simulations for their students, highlighting the need for an understanding of Science, Technology, Engineering and Mathematics (STEM) disciplines and encouraging them to pursue STEM-type careers.

Figure 92—Simulated plane about to crash into building
(Intellectual Property of Dr. Hoffman)

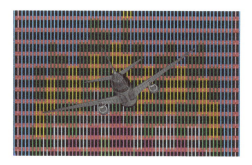

Figure 93—Simulated plane at impact
(Intellectual Property of Dr. Hoffman)

Medical

My first exposure to medical simulation was sometime, I believe, in the 1980s. I was with the Navy and was asked to participate in reviewing technology proposals for the State of Florida. At that time, the state set aside a limited amount of funds to support research and development in areas that showed promise. Members from industry and UCF participated in the review, as I recall. One of the proposals was from Dr.Michael Good with the

University of Florida (UF) at Gainesville for a Human Patient Simulator. There was much debate at the time with opponents not convinced that the human body could be simulated and even if it could to some degree, how would it be used? He received funding and went on to get additional funds the following years. To me that was the beginning of what would be the "new frontier" of modeling and simulation-medical and health care. Since that time, others have entered the field. One such individual is Dr. Richard Satava of the University of Washington. I was pleased to meet Dr. Satava a long time ago at I/ITSEC. He is one of the medical simulation pioneers and served on the White House Office of Science and Technology Policy.

At the University of Florida, the Center for Safety, Simulation & Advanced Learning Technologies (CSSALT) provides education, training and services to residents, faculty, clinical personnel and staff throughout the UF Academic Health Center including the Shands Healthcare system, to clinicians in the state of Florida, to UF medical, veterinary, dental, engineering and other health profession students, to local and regional emergency personnel and to industry executives and personnel worldwide. CSSALT builds on the legacy of a continuous and sustained R&D effort in simulation that began in 1985 with the "Bain Group" under the direction of Drs. Gravenstein and Beneken and remains active today in the development and application of simulation technology. The current dean of the UF College of Medicine, Dr. Good and the director of CSSALT, Dr. Lampotang, developed in 1987 the first prototype of what would become the UF Human Patient Simulator (HPS) mannequin patient simulator technology that is licensed to Medical Education Technologies, Inc. (METI) and used worldwide. (Figure 94)

The HPS has heartbeat, pulses, breathes, blinks his/her eyes and reacts to drugs much like a human patient. It is used by colleges and universities and in technical schools throughout the world. CAE Healthcare with patented cardiovascular, respiratory and neurological and pharmacological modeling has a HPS with true oxygen and CO_2 gas exchange, the HPS can also rise to the challenge of accurately representing complex surgical, critical care and drug interaction scenarios.

The Virtual Anesthesia Machine (VAM) team, affiliated with CSSALT, offers a portfolio of web-enabled simulations at **http://vam.anest.ufl.edu/wip. html.** *(UF website)*

Figure 94—A Medical Simulation Laboratory with Human Patient Simulator

Nicholson Center

Training thousands of physicians worldwide, the Florida Hospital Nicholson Center at Celebration, a stone's throw from Disney in Celebration, FL, leads the way in developing, incubating and accelerating next generation clinical knowledge, technology and treatments for the global community. (Figure 95)

Figure 95—Florida Hospital Nicholson Center

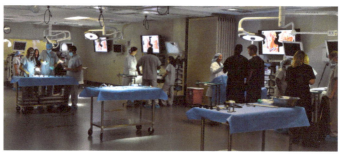

The 54,000 square foot facility in Central Florida features state-of-the-art surgical suites and simulation labs, giving physicians and surgeons hands-on experience with current and emerging technologies. A multi-specialty robotic training lab and 25-station surgical skills lab allows physicians to perform complex procedures in a simulated environment that reproduces real world situations.

A Center of Excellence in medical robotic and surgical simulation, is not only able to provide physicians on-site surgical training using the latest technologies, but can bring in other professionals from around the world via live broadcasts, two-way video conferencing, telementoring and web-based education.

The Nicholson Center features:

- A 500-seat conference center with high definition, multi-campus site videoconferencing and global communication capabilities.
- 6 da Vinci robotic stations in a dedicated robotics lab with bidirectional connectivity to other learning and education spaces. (Figure 96)
- A simulation and robotics training center that offers computer-based simulations, medical virtual reality trainers, surgical simulators and multiple robotic technologies.
- An innovation and technology accelerator that focuses on new business initiatives with physicians, global medical companies, medical associations and the military.
- A dedicated dry simulation and robotics lab that can be subdivided into three suites, each offering full AV integration.
- Two 935 square foot, fully equipped team training operating room with bidirectional AV integration.
- Conference rooms, collaboration suites and board rooms accommodating teams from 4 to 25.

*Figure 96—DaVinci Surgical System
(Intuitive Surgical Inc.)*

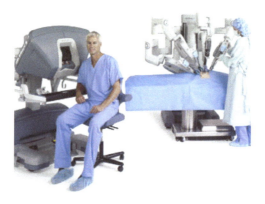

CAESAR Trauma Simulator

Built for trauma, disaster response and combat casualty care, Caesar (Figure 97) is the most rugged patient simulator available today. With life-sized realism and modeled physiology, Caesar offers clinical accuracy for basic to advanced point-of-injury training. Deploy Caesar to any challenging climate, terrain or training environment. Through tourniquet placements, patient decontamination, and extreme temperatures and conditions, Caesar remains tough-skinned and resilient.

Caesar is built for durability in order to train emergency field responders who may be subjected to harsh climates and terrains. With rugged skin that is water-resistant, Caesar can rise to the challenge of demanding, point-of-care training. For added realism, Caesar has articulation of his neck, back, shoulder, elbows, forearms and wrists as well as directional eye movement and speech patterns that reflect his level of consciousness.

With the capacity to hold up to 1.4 liters of "blood" on board, Caesar can present dramatic bleeding from up to six sites and produce automatic physiological responses to tourniquet application. Caesar is built with modular limbs. His optional wounds kit includes right hand glove with gunshot wound, abdominal multiple gunshot wound, wrist injury and trauma face. *(CAE Healthcare website)*

Figure 97—CAESAR Trauma Simulator (US Army)

In addition to Caesar, there is a wide spectrum of medical simulators, and the list continues to grow as more people discover the potential of simulation. The following is a list of examples of common medical simulators used for training.

- Advanced Cardiac Life Support simulators
- Partial Human Patient Simulator (Low tech)
- Human Patient Simulator (High tech)
- Hands-on Suture Simulator (Low tech)
- IV Trainer to Augment Human Patient Simulator (Low tech)
- Pure Software Simulation (High tech)
- Anesthesiology Simulator (High tech)
- Minimally Invasive Surgery Trainer (High tech)
- Bronchoscopy Simulator
- Battlefield Trauma to Augment Human Patient Simulator
- Team Training Suite
- "Harvey" mannequin (Low tech)

Advantages

Studies have shown that students perform better and have higher retention rates than colleagues under strict traditional methods of medical training. The table below shows the results of tests given to 20 students using highly advanced medical simulation training materials and others given traditional paper based tests. It was found that high technology learning students outperformed traditional students significantly.

E-Learning vs. Textbook Learning		
Mode of Learning	Mean Test Score on Multiple Choice Test	Time to Complete Module
E-Learning (N=20)	4.03 / 5 (80.6%) "B"	28–30 minutes
Traditional Paper Based	3.05 / 5 (61%) "D"	28–30 minutes
Significant Difference	Yes (p <.001)	N/A

In addition to overall better scores for medical students, several other distinct advantages exist not specifically related to training.

- Less costly
- Time efficient
- Less personnel required
- Many automated processes
- Ability to store performance history
- Track global statistics for many linked medical simulators
- Less medical related accidents

Military and emergency response

One of the single largest proponents behind simulators has always been the United States Government. Billions (and perhaps trillions, at this point) of dollars have been spent in the name of advancing simulators for space exploration, computer advancements, medical and military training, and other projects funded for research by the government. The DoD (most notably, the Army) is one of the largest sources of funds for simulation research, training, and support. As such, most simulators tend to be created for military purposes including soldier, tank, and flight training in combat situations. In terms of medical simulation, military applications have played a large part in its success and funding. Some examples of scenarios useful for medical applications include casualty assessment, war trauma response, emergency evacuations, training for communications between teams, team/individual after action assessment, and scenario recreation from recorded data. (*Wikipedia*).

Medical City

One of the fastest growing medical complexes in the nation is at Lake Nona adjacent to Orlando, Florida. They are calling the complex "Medical City".

It's been called a new chapter in Orlando's history. The 650-acre health and life sciences park known as Lake Nona Medical City is a landmark for Orlando and a premier location for medical care, research and education. Carefully planned and laid out, Lake Nona Medical City represents a deliberate strategy to create a centralized focus of sophisticated medical treatment, research and education in Central Florida.

Based on the proven theory that a cluster of healthcare and bioscience facilities in proximity to one another will accelerate innovation, this intellectual hub opened in a coordinated fashion with a collaborative mission. In the next decade, Lake Nona Medical City will be home to some of the nation's top hospitals, universities, research institutions and life science companies. But already, the Medical City's pioneering institutions are forming networks and synergies making Orlando a global destination for healthcare, research and medical education while creating an economic development and job creation engine for the region. (*Learnlakenona.com*)

Medical City is anchored by the following: (Figure 98).

- University of Central Florida Health Sciences Campus
- Sanford-Burnham Medical Research Institute
- Veterans Administration Medical Center (Serving 400,000 veterans)
- Nemours Children's Hospital
- University of Florida Academic and Research Center
- M.D. Anderson Orlando Cancer Research Institute

Figure 98—Florida's Medical City near Orlando (Okraski)

Burnham Institute for Medical Research

Veterans Affairs Medical Center

University of Central Florida/Burnett

University of Florida

Nemours Children's Hospital

M.D. Anderson Cancer Research

Entertainment

People have been entertained by various forms of abstraction for thousands of years. Simulation followed drawings, comic books, plays and movies. Location-based entertainment centers and theme parks included games and various kinds of simulation. One can take his or her children to a pizza center where they can ride a simulated motorcycle by dropping a few coins in a slot. They can also visit the epicenter of entertainment simulation in Orlando with its Disney World, Universal, Sea World, etc. It is ironic that the Disney Corporation selected Central Florida to construct "Disney World", rich with animatronics and other simulation technologies- all aimed at suspending disbelief...a place where one can escape from the cares of the real world.

If you recall, the Navy, Army and Marines were already in Orlando developing simulation for the military. Not that there was any dialogue between the two developers of simulation technology. Disney, at that time, did not want to be associated with folks that prepare for war. It wasn't good public relations. However, today there is more conversation between the entertainment community (the west side of Orlando) and the military (the east side of town). Further, this is true on the national stage.

Out at Kennedy Space Center, NASA has recently expanded their Visitors Center to include a vast array of models and simulations. The $100 million investment is impressive. There are some simple physical models (Figure 99) and more interactive simulations.(Figure 100). The space shuttle Atlantis has been spruced-up and is now center stage in a vast array in a very riveting production appealing to all ages. To support the shuttle, you can experience several simulations including docking and other tasks the crew would perform on a mission.

Figure 99—Shuttle Physical Model at
Kennedy Space Center Visitors Center (Okraski)

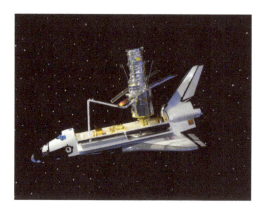

Figure 100—Part Task Simulator at
Kennedy Space Center Visitors Center (Okraski)

Perhaps the most exciting experience is the launch simulator. Here, several of you are pre-briefed on the launch, its various stages and what to expect in the ride. Videos of astronauts are shown attesting to the accuracy of the simulation when compared to actual flight. Strapped in, seats rotated, you are ready for an eight and one half minute journey into orbit. Then it is "blast-off" with all the vibration and sound you might expect. It is a fun ride with good simulation and will surely generate interest in the space program.

In 1996, representatives from the entertainment and Department of Defense research communities met under the aegis of the Computer Science and Telecommunications Board of the National Research Council to discuss "Linking Entertainment and Defense".(National Academy Press). The workshop was chaired by Dr. Michael Zyda, Naval Postgraduate School. The group concluded that there are areas of mutual interest but also recognize cultural differences. Research areas of common interest were identified as follows:

- Technologies for Immersion
- Networked Simulation
- Standards for Interoperability
- Computer-generated Characters
- Tools for Creating Simulated Environments

At present, the same contractors that develop simulators for the military develop systems for the entertainment community. Visual displays, computer image generation, motion platforms, 3D viewing and even olfactory simulation are dual use.

One important point, however, is that training simulators are designed differently than entertainment venues. Training simulators are designed to meet identifyable behavioral objectives whereas simulators for entertainment are designed to provide a thrill with plenty of bumps and grinds. One reason the duration of the ride is short is because the designers do not want the customer to get sick And sick they will get when the visual cues do not agree with what the body feels. "Simulator sickness" is what we call it in the trade.

Entertainment and education can often be combined into what some call "edutainment". We know children like being entertained, particularly if they are actively involved. One good example is found at the Denver Childrens' Museum where a fire truck and its functions are used to teach children how to call "911" in the event of a fire, how it feels to be a firefighter, how they live, etc. (Figures 101 and 102).

Figure 101—Simulated Fire Engine (H.Okraski)

Figure 102—Young Firefighter at the simulated fire truck controls (My Grandson) (H.Okraski)

The latest appealing technology involves the creation of avatars and creating situations where avatars represent real players. Programs are created using commercially available services and software and are employed in entertainment, business, military, education and others. (Figure 103). It seems that students can be both entertained by the technology and learn from the unique experience. Expect to see more of this captivating technology in the future.

Figure 103—Virtual World using Avatars

Commercial Airlines

The first transfer of simulation technology from the military, if we consider the Link Trainer as the first "military trainer technology", to civilian use was to the commercial airlines. In 1937, because there were no other trainers readily available, the airlines began to order Link Trainers. American Airlines purchased a Model "C" to become the first commercial customer for a trainer. Immediately following that, United Airlines, Pan American, Australia National Airlines, Eastern Airlines and BOAC also purchased Link Trainers. These trainers were valuable in teaching radio compass. At that time, there was fierce competition between Link and Curtiss-Wright. There was even a long legal suit between the two who dominated the limited, but promising commercial simulator market. However, it appeared that the most feasible approach to satisfying market demands was for Curtiss-Wright to pursue propeller-driven aircraft and Link the jets.(*Kelly*)

However, along came the first commercial jetliner, the DC-8. Curtiss-Wright and Link went head-to-head in the competition. Curtiss was awarded a contract by Pan American for a DC-8 Simulator and Link a contract for the delivery of two simulators, one each to United Airlines and Douglas. The Link simulator was the first simulator to be delivered prior to the delivery of the aircraft. (This is a condition the military has tried to equal over the years with some success. The A-7A Weapons System Trainer was one of them.)

Ray L. Page has chronicled the history of commercial simulators as follows:

"Dr. Dehmel from the Bell Laboratories had continued with his interest in simulation and through his work on Bell's M-9 anti-aircraft gun directors, applied this knowledge to the design of an instrument flight trainer and in 1943 was able to convince the Curtiss-Wright Corporation to manufacture these devices. Curtiss- Wright continued their interest after the war and contracted to Pan American Airways to construct a full simulator for the Boeing 377 Stratocruiser which became the first full aircraft simulator to be owned by an airline. (Figure 104).

In 1947, BOAC decided to buy Boeing 377 Stratocruisers and knowing of Redifon's work on synthetic trainers, requested a proposal from this company to construct a device similar to that being used by Pan American. This resulted in an agreement being entered into between Redifon and Curtiss-Wright with work commencing in 1950.

These devices used the alternating current. carrier method of analogue computation with contoured potentiometers and 400 Hz synchros for aircraft instrument drives. The control loading unit used variable levers, servo controlled as a correctly computed function of airspeed with springs to provide the required forces. This simulator was completed in 1951 at a cost of £120,000. Prior to completion of this simulator, Redifon gained another contract from BOAC for the Comet 1 which was to become the first jet transport simulator to be constructed. A number of airlines throughout the world purchased REDIFON simulators including QANTAS.

Figure 104—Curtiss-Wright Commercial Airline Simulator
(Ray Page)

After the war, competition from Curtiss-Wright stimulated Link to de-velop their own electronic simulators using analogue computation and this was used in their C-11 Jet Trainer for which a contract was awarded by the U.S. Air Force in 1949. Over a thousand of these were produced. Link moved from alternating current to direct current for the analogue computation which was a far more demanding technology but capable of far greater precision, and in the mid 1950's, Pan American and QA-NTAS became the first airlines to place into service simulators for the B707 aircraft which at that time would have then been the world's most sophisticated simulators ever produced "*(Ray Page)*

Over the years, commercial airline simulators have increased in fidelity es-pecially in motion and visual systems. (Figure 105)

Figure 105—Airbus simulator

Airlines rely heavily on the use of simulators in training pilots and air crews. (Figures106 and 107). Some commercial airlines have their own training centers while many pay for training at a facility that provides those services to American and international customers. Commercial simulators have proven to be effective in substituting for aircraft time. They save money, time and lives just like military simulators. However in the business world, return on investment (ROI) is measured in dollars and cents. The "unusual" can be practiced in a simulator whereas it may be unsafe or impractical to practice in the actual aircraft. Simulators become vital when it comes to experiencing the low probability but high impact occurrences. Undoubtedly, the unparrelled aviation safety record of the US airlines is, in part, attributed to the use of simulators. Evidence of the confidence placed on simulator training is the fact that pilots can be transitioned to aircraft without training flights in the actual aircraft. The Federal Aviation Administration (FAA) sets very high standards for the systems simulated and degree of fidelity in the simulation. In order for the simulator to be certified, certain criteria must be met. Simulators and other training devices are catagorized by type and appropriate acceptance criteria applied to each category. For example, a full-up flight simulator will have to meet Level D criteria for training. The general requirements (summarized) are shown in Table 1.

Table 1 —US Federal Aviation Administration (FAA) Simulator Requirements

Flight Training Devices (FTD)

- **FAA FTD Level 4**—Similar to a Cockpit Procedures Trainer (CPT), but for helicopters only. This level does not require an aerodynamic model, but accurate systems modeling is required.
- **FAA FTD Level 5**—Aerodynamic programming and systems modeling is required, but it may represent a family of aircraft rather than only one specific model.
- **FAA FTD Level 6**—Aircraft-model-specific aerodynamic programming, control feel, and physical cockpit are required.
- **FAA FTD Level 7**—Model specific, helicopter only. All applicable aerodynamics, flight controls, and systems must be modeled. A vibration system must be supplied. This is the first level to require a visual system.

Full Flight Simulators (FFS) **FAA FFS Level A**—A motion system is required with at least three degrees of freedom. Airplanes only.

- **FAA FFS Level B**—Requires three axis motion and a higher-fidelity aerodynamic model than does Level A. The lowest level of helicopter flight simulator.
- **FAA FFS Level C**—Requires a motion platform with all six degrees of freedom. Also lower transport delay (latency) over levels A & B. The visual system must have an outside-world horizontal field of view of at least 75 degrees for each pilot.
- **FAA FFS Level D**—The highest level of FFS qualification currently available. Requirements are for Level C with additions. The motion platform must have all six degrees of freedom, and the visual system must have an outside-world horizontal field of view of at least 150 degrees, with a collimated (distant focus) display. Realistic sounds in the cockpit are required, as well as a number of special motion and visual effects. *(Wikipedia)*

Figure 106—Thales flight simulator at a pitch angle

Figure 107—Twin jet simulator cockpit

Education

With an ever increasing emphasis on STEM (Science, Technology, Engineering and Mathematics) we will be experiencing more simulation technology in the classroom. With few exceptions, teachers of today are not familiar with simulation, how the technology can be used in the classroom to demonstrate concepts and, perhaps most importantly, what careers are available in simulation. Perhaps one of the most significant obstacles to introducing new technologies into the classroom is the total focus on competency exams. Teachers have commented that simulation is nice but we have to teach to the exam.

There are, however, pockets throughout the country where STEM/M&S is gaining a foothold. The Tidewater area in Virginia (Old Dominion University) is one and Huntsville, Alabama another. There are M&S courses being given in the Wright-Patterson AFB region in Ohio in Montgomery County. The Central Florida High Technology Corridor has several initiatives underway. I can speak with some authority in that regard because I recently chaired a task force, under the aegis of the National Center for Simulation, to develop a high school program in modeling and simulation. We have at least six high schools in the Central Florida area that have introduced simulation into the classroom. Finding or developing teachers to teach the material is a serious issue and will take a lot of work to put a responsive system in place.

At the University of Central Florida (UCF), education professor Dr. Lisa Dieker, computer sciences professor Dr. Charles Hughes and mathematics professor Dr. Michael Hynes worked together for five years to develop a simulated classroom called TeachLive. (Figure 108). UCF, including its Institute for Simulation and Training, is a leader in simulation and training throughout the nation. With TeachLive, a student teacher can experience and practice the interaction with a variation of student personalities, attitudes and abilities. Just as with flight simulators, student teachers can develop the skills that will enable them to perform in their school environment after graduation. The system actually generates avatars that represent the cross section of students. Live actors play the parts of the students. By exposing student teachers to children with various qualities, the teacher develops confidence and with confidence comes competence. TeachLive is just the beginning. Eventually, the system could have artificial intelligence replace the actors and there could be some "coaching" incorporated into the system. Perhaps there would be some performance measurement incorporated in the future—which looks very promising.

Figure 108—TeachLive Simulation at UCF
(Intellectual Property of the University of Central Florida)

One of the most gratifying education projects that I have undertaken was a virtual reality (VR) initiative for deaf and hearing impaired children. It began with the principal of an elementary school attending an NCS Board of Directors meeting. He said he needed help. The problem he outlined was that teachers spend most of their time in working with the children in developing life skills and have little time for course work. I was working with Veridian, Inc. as a consultant and we put together a proposal for the US Department of Education which involved the Schools for the Deaf and Blind in Florida, Pennsylvania and Ohio.We had the support of Congressman Bill McCullum of Florida and the legislators of the other two states. The project was funded and we delivered a VR program with a head-mounted display that had the children practice crossing the street, ordering at a fast food restaurant, danger stranger, how to evacuate the school in the evnt of a fire, how to dial 911, etc. (Figures 109 and 110). The children loved it and then we began to introduce some math into the scenarios. The principal was a strong supporter and we had a "teacher-champion" in the school that was necessary to make it a success. An evaluation by UCF's IST validated its acceptance and performance.

Figure 109—Virtual Environment for Deaf and Hearing Impaired Children (Veridian, Inc.)

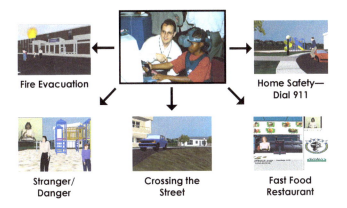

Fire Evacuation

Home Safety—
Dial 911

Stranger/
Danger

Crossing the
Street

Fast Food
Restaurant

Figure 110—Author with Immersed Student (H.Okraski)

Transportation
Analysis

Modeling and simulation are applied to flight training, commercial airlines, marine operations and other forms of transportation. Analytic models are used by the various departments of transportation in planning of roads, bridges, etc. Simulations provide visualizations that help decision-makers in understanding the impacts of alternatives being considered. (Figure 111). For example, if options are being considered that include adding more toll booths to an expressway, the effects of that option can be analyzed through modeling and simulation. The same is true for road widening or adding new roads. The options can be studied from the aspects of revenue, traffic flow, safety, etc. In the past, results were presented as computer print-outs. Today, the data can be presented in visual format. So much more information can readily be presented to the decision-maker in lieu of reams of paper with statistical data that are difficult to sort through.

Figure 111—Transportation Management using simulation

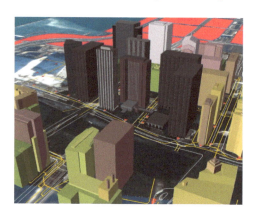

Driver Trainers

Driver trainers for the general public and high schools have been around for a long time. As a teenager at Utica Free Academy I know my driver training instructor would have welcomed a simulator to practice before braving the road with a heavy-footed teenager. I remember, vividly, the look of horror on his face when a student would attempt a left turn in front of an oncoming truck. His life was shortened each and every day. In many schools "live" driving in driver education has been discontinued because of liability issues. In those schools, they rely totally on driver trainers for practical driving, and the parents have to use the family car for "live driving".

The fidelity of driving simulators for automobiles, trains, trucks, etc. has gotten much better with improved visuals and motion simulation. Motion adds more realisim. Without motion there can be a tendency to get "simulator sickness" in these trainers.

The military uses driver trainers to a large degree not only for individual skills training but for operating a vehicle in a convoy. Experience has shown that convoy training is needed for situational awareness, route planning and tactics in driving in hostile areas. The Jessica Lynch situation underscored the need for this kind of expanded training.

Virtual Combat Convoy Trainer (VCCT)

Shortly after the Jessica Lynch incident, there was an urgent requirement to deliver convoy trainers to the Army. One such trainer is the Virtual Combat Convoy Trainer (VCCT). (Figure 112).

Figure 112—Virtual Combat Convoy Trainer
(Lockheed Martin and Firearms Training Systems)

The VCCT enables combat crews to communicate, maintain situational awareness and acquire targets while moving at highway speeds operating in a convoy environment. The trainers are expected to improve convoy tactics and minimize combat related injuries and deaths resulting from attacks on convoys. One-third of all US casualties since the beginning of Operation Iraqi Freedom were caused by attacks and accidents related to convoy traffic.

Common Driver Trainer (CDT)

The Common Driver Trainer (CDT) system consists of virtual trainers, which will be used to train members of the armed forces in the operation of the combat vehicles being fielded to Fort Leonard Wood, Mo and to bases in both Korea and Japan, as well as technical refresh to existing CDT systems fielded through the US. (Figure 113).

The CDT consists of a simulated vehicle cab, instructor/operator station, After Action Review (AAR) station, visual system, six-degrees-of-freedom motion system and a computational system. Via the instructor/operator station, the instructor is capable of selecting a visual scene, introducing malfunctions and emergency control situations, monitoring each Soldier's performance and providing recorded AAR feedback. The reconfigurable common platform provides driver training for U.S. Army tactical vehicles including the M1A2 Abrams, Stryker, Mine Resistant, Ambush Protected (MRAP) vehicle, MRAP All Terrain Vehicle (MATV) and the Abrams Tank Engineering Variant (TEV). The CDT has been fielded in both fixed-site and mobile configurations.

Figure 113—Common Driver Trainer (MRAP Variant) (PEO STRI)

National Advanced Driving Simulator (NADS)

The top of the line of civilian driving simulators is the National Advanced Driving Simulator (NADS) located in Iowa. (Figure 114)).

Figure 114—NADS Simulator

NADS is the most sophisticated research driving simulator in the world. Developed by the National Highway Traffic Safety Administration, the NADS offers high-fidelity, real-time driving simulation. It consists of a large dome in which entire cars and the cabs of trucks and buses can be mounted. The vehicle cabs are equipped electronically and mechanically using instrumentation specific to their make and model. At the same time, the motion system, on which the dome is mounted, will provide 400 square meters of horizontal and longitudinal travel and nearly 360 degrees of rotation in either direction. The effect will be that the driver will feel acceleration, braking and steering cues as if he or she were actually driving a real car, truck or bus.

The latest in visual display technology and a high-fidelity audio system will complete the driving experience. The test subject will be immersed in sight, sound and movement so real that impending crash scenarios can be convincingly presented with no danger to the subject. Vehicle and driver data can be accurately gathered and stored and tests repeated with exactitude. TRW was awarded the construction contract after winning the design competition conducted by NHTSA. TRW assembled a world-class team that incorporated the latest in simulation technology in all aspects of the design for the $50 million project. The team included Dynamic Research, Inc., Evans & Sutherland, I*SIM and MTS Systems.

The NADS is located at the University of Iowa's Oakdale Research Park, Iowa City, IA. The site was selected as a result of a national competition among major transportation research universities conducted for NHTSA by the National Science Foundation. The University of Iowa provided $11.58 million in cost sharing to the NADS project, which included the design and construction of a $5.7 million building to house the simulator.

The effects of alcohol, drugs, visual impairments and aging on driving will all be safely studied using the new research tool. The NADS will provide the capability for evaluating advanced vehicle communication, navigation and control technologies which are now being developed as part of the Intelligent Transportation System (ITS) program. The medical community will look to the NADS to answer questions about the development and effects of new medicines and prosthetics on driving. Car companies can use the NADS to develop and test new safety devices, in conjunction with their own simulators. Human factors issues, estimated to be a contributing cause of 90 percent of motor vehicle accidents, will finally be able to be studied in a safe, accurate and repeatable environment. (Wikipedia)

Rail Simulators

The general public may not be aware that there is a great deal of research conducted on our nation's railway system. Much of this research is conducted at the Volpe National Transportation Systems Center, part of the U.S. Department of Transportation's Research and Innovative Technology Administration. Volpe is a critical resource for innovation in transportation. Their mission is to improve the nation's transportation system by anticipating emerging transportation issues and to serve as a center of excellence for informed decision making. They use simulation quite extensively. One example of a human factors simulation is the Cab Technology Integration Laboratory (CTIL).(Figure 115).

The CTIL is a state-of-the-art, full-sized locomotive simulator that allows researchers to simulate a number of conditions and scenarios encountered during railroad operations. From these simulations safety problems can be identified, analyzed and effective solutions developed without having to attempt to reproduce certain conditions in the real world. In fact, it may be impossible to create dangerous conditions in reality.

The full-sized locomotive simulator can accurately record crew behavior through the use of video, audio, and even eye-tracking capabilities. This allows researchers to carefully observe the actions of train crews and monitor the corresponding effect of their actions on the simulated locomotive they are operating.

Other features include modeling and visualization technologies that allow the laboratory to work with locomotive designers to assess how new technologies and control configurations affect operator performance.

Figure 115—Rail Simulator at Volpe National Transportation System Center (Wikipedia)

While on the topic of railway transportation few are aware that military trainers were mobile at one time. They were trailerized and the Air Force even moved their B-52 Simulators to the Strategic Air Command (SAC) bases by rail.(Figures 116, 117 and 118). *http//fastlane.dot.gov*

Figure 116—B-52 Simulator by Railroad Car (Air Force)

Figure 117—B-52 Simulator Inside the Rail Car (Air Force)

Figure 118—Pilot in more recent B-52 Simulator
(www.airforce-technology.com)

Maritime and Ship Simulators

There is an interesting history of ship simulators. Although crude simulators were developed by Antisubmarine Warfare Operations Research Group
and by the Naval Ordnance Laboratory during WWII, true ship simulators
began to appear in early 1945. The University of California developed a limited simulator, emulating a ship's combat information center. Known as the
Naval Electronic Warfare Simulator, the idea came to the attention of the
Chief of Naval Operations, Fleet Admiral Ernest J. King, who directed the
development of the first naval surface warfare simulators.

In 1952, RAND Corporation developed the first true ship simulators. The
Systems Research Laboratory was built by a team of behavioral scientists to
investigate the optimal training methods for improving group performance.

By 1994, ship simulators were becoming more of a teaching tool. Simulators
have been installed at the U.S. Merchant Marine Academy, several private
maritime schools and at the U.S. Naval Academy

Virtual ship simulators include bridge simulators, part task trainers and
weapons systems trainers that appear to be the "real thing ". There are also
maintenance trainers to teach troubleshooting of specific equipment. Bridge
simulators are used by prospective deck officers-those who will control the
navigational operations of a ship and are employed to train mariners to handle ships in a variety of situations, from docking and undocking, to navigat-
ing various approaches in a variety of conditions using actual ship performance data in a real time. NAWCTSD has developed a Submarine Piloting
and Navigation Trainer (SPAN) that uses voice recognition and artificial
intelligence to enable safe passage, when surfaced, through narrow channels
in the virtual world. Part task trainers provide a training platform to teach
specific equipment found on-board the ship. On- board military ship simulators, crews also can train as teams that might include radar, fire control,
electronic warfare, anti-submarine warfare, combat systems, etc.

For the merchant marine and other non-military ships, the tasks are less complex. However, with the introduction of terrorisim and pirate activities emerging on the high seas, the need for combat training is more evident..

Figure 119—Simulated Bridge (http://wikimedia.com)

The captain (Figures 119 and 120) keeps an eye on a cargo ship's bridge controls during a demonstration of how they use simulators at the Simulation, Training, Assessment & Research Center (STAR) to teach merchant marines, among many things, how to react to a potential threat from pirates. Most of what they teach now is non-lethal, such as maneuvering the boat or the use of water canon to thwart a boarding attempt. STAR trains shipping companies throughout the world, including Maersk, whose cargo ship Alabama was overtaken by pirates near Somalia's coast en route to the Kenyan port of Mombasa. (Photo by Joe Raedle/Getty Images)

Figure 120—Captain at Bridge Controls in the simulator

(Joe Raedle/Getty Images North America))

Military Air Traffic Control Simulation

In the very early years of air traffic control and carrier air traffic control training, scaled airplane models with humans holding them were used to simulate aircraft. Beyond that era, I recall working at the FAA experimental center near Atlantic City where controllers were trained using very primitive radar target generators. There were rows of desk top simulators in a room with operators positioned at consoles. Each position simulated a particular aircraft in the landing pattern. All of these were connected to a mainframe computer that was very large. Operators were part time employees who had to learn the language of air traffic conversation because they were simulating pilots in the scenarios.

At the Navy, we had difficulty with early simulators in the speech recognition and the intelligence simulated. First of all, the speech recognition system was not speaker independent. Secondly, any deviations from the standard conversation protocols put the trainer in a tailspin. Fortunately, technology has matured to where those same difficulties are not issues.

The Navy has deployed an air traffic control tower simulator system (ATC-TSS) to the Naval Air Station Key West, Florida. (Figure 121) The US Navy is planning to procure a total of 38 simulators to provide low-cost proficiency training for its naval air traffic controllers, while replacing its existing Tower Operating Training System (TOTS) at 34 military facilities. Comprising built-in scenarios such as emergency situations and daily routines, the new simulator system will overcome the TOT's issues such as decreased visual capabilities and speech recognition programs. Supporting individual or team training, the ATC-TSS provides both out-of-the-window and binocular views as well as 3D graphics with simulated weather information, airfield lighting and integrated radar displays.

Figure 121—Navy Air Traffic Control Simulator

Photo: courtesy of US Navy

Civilian Air Traffic Control Tower Simulators

An air traffic controller (Figure122) works inside an airport tower simulator at the Denver International Airport in Denver, Colorado. The Federal Aviation Administration (FAA), has passed 75 years of federal air traffic control in the United States. Beginning with just 15 employees in 1936, the FAA now has some 15,000 air traffic controllers and technicians working nationwide. With that volume of controllers and the need to train new employees and refresh current personnel, simulation is the only alternative. Plus, it is safe and cost effective. When a controller makes a mistake in the simulator he or she learns from the experience-rather than recover from it. *(www.zimbio.com)*

Figure 122—Air Traffic Controller at Denver International Airport Simulator

(John Moore/Getty Images North America)

Homeland Security and Public Safety

The need for simulation was driven home loudly as a result of 9/11. This tragedy underscored the need for more coordinated training of various law enforcement agencies in disaster management. Simulation is a very effective tool in training teams how to respond to disasters of various types. Simulation is used by the military in mission planning and mission rehearsal and that same technology can be used for civilian purposes.

Shortly after 9/11, I was invited by Congressman John Mica (R-FL) to go before a Congressional Select Committee to advise the members on the effectiveness of simulation in planning for responses to disasters and in execution by first responders and others. Live, virtual and constructive simulations will ensure that our homeland security forces are operationally ready when disaster occurs and that predetermined and pre-trained plans can be executed.

What made the 9/11 particularly disturbing to us from the simulation community is that the terrorists also appreciated the value of simulation and practiced the attack in a flight simulator. We have to wake up and control who we provide training to by doing background checks, developing personality assessments and other methods that could be used by the FBI and others. Further, a data base of pilots and those in pilot training should be kept and reviewed by our law enforcement agencies. We have data bases for gun control, automobile owners, etc., surely the damage that can be incurred with an airplane is several magnitudes beyond.

In the past, the military put a lot of emphasis on individual training but the USS Stark incident in 1987 in the Iran-Iraq War and the USS Vincennes incident where an Iranian civilian jet airliner was shot down by U.S. missiles sparked an interest in team training, specifically tactical decision-making under stress. Two researchers at the Naval Air Warfare Center Training Systems Division, Drs. Eduardo Salas and Jan Cannon-Bowers worked with the military to develop methods for that kind of training. It applies not only to military teams but also to civilian and civilian/military teams involved in disaster management of all kinds.

Today, there are several disaster management simulations available such as the Advanced Disaster Management Simulator (ADMS). (Figure 123) and other virtual simulations that deal with security in various settings such as the Capitol Hill Police Checkpoint Trainer (Figure 124).At the aggregate level there are many constructive simulations used at the local, state and national levels.

Figure 123—Advanced Disaster Management Simulator (ADMS)

ADMS trains all phases of on-scene incident command, training incident commanders, command post staff, vehicle operators and others. It can offer a team training solution for the entire chain of command or individuals.

Figure 124—The Capitol Hill Police Checkpoint Trainer

The Capitol Hill Police Checkpoint Trainer uses complex 3D simulations to emphasize proper security deployment and response procedures, the principles of access control and perimeter security, bomb threat and search procedures, and proper reporting and communication etiquette. Scenarios are comprised of different types of critical checkpoint incidents and their effects on the population and transportation infra-structure surrounding the Capitol complex. The Checkpoint Trainer requires users to screen all vehicles. The extent of the screening depends on the type of vehicle, what the vehicle appears to be carrying and how easy or difficult it is for officers to look inside. *(ECS Website)*

A few years ago, I collaborated with an Orlando company to develop a Fire-fighter's Personal Trainer. (Figure125) We submitted a grant request but it was not accepted by the sponsor in spite of the fact that all of the firefighters we met with loved it. The idea is to use a virtual simulation of various fires in a more real approach. The firefighter would have to move and maneuver according to the scenario being presented outfitted in normal fire fighting gear. There would be simulated communications between the exercise director, the firefighter and, if desired, another firefighter networked to the first.. In other words, he or she would get a good workout without leaving the fire station. Feedback could be provided regarding procedures, safety, tactics, etc. Fire stations could compete with one another. Although the proposal was not accepted, I have learned in this business that it pays to be persistent if you have a good idea. Timing is everything!

Figure 125—Fire fighters Personal Trainer Concept
(H.Okraski and Engineering & Computer Simulations)

Nuclear Power Plant (NPP) Simulators

Simulators are ideal for teaching the low probability but high impact tasks. For example, with aircraft crew training it is not too likely they would flame-out all engines of the airplane just to see if the air crew knew how to handle the simultaneous failures. That is where the simulator becomes valuable. It doesn't matter if the simulated plane crashes and burns. In the actual aircraft it becomes another story. The same is true with nuclear power plants and their simulators. In the simulator, both high probability and low probability failures can be inserted for the maintenance crew to practice. Nuclear power plant simulators must be high fidelity in both physical and functional design. One of the comments I have heard concerning NPP simulators is that they are no different than our flight simulators-often the configuration is not consistent with the parent system. This is the nemesis of the simulator world in general but with NPP simulators one mistake or bit of negative training can kill thousands or more people and destroy the lives of many more. What you see in the simulator is what you hope to have in the NPP itself. Figures 126 and 127 are representative power plant simulators.

Figure 126—European Nuclear Power Plant Simulator

Figure 127—United States Nuclear Power Plant Simulator

The names that come to mind when discussing nuclear power plants are Three Mile Island (1979), Chernobyl (1986) and Fukushima (2011). In the first two incidents, human error was determined to be a significant factor in the accidents. The earthquake of magnitude 9.0, tsunami and tidal water were the primary causes of the Fukushima disaster which ranked very high on the nuclear disaster scale. Chernobyl and Fufushima were the two worst nuclear power accidents in history.

After Three Mile Island in 1979, there was a renewed and heightened interest in the use of simulators for training. In fact, the following trends were taking shape:

1) Greater use of control room simulators
2) More specific requirements by regulatory bodies concerning simulator training
3) Greater emphasis on soft skills
4) Considering simulator training as an integral part of overall training program for specific jobs
5) Better incorporation of operating experience into simulator training programs and materials
6) Use of simulators for training a variety of NPP jobs
7) Better methods to assess individual and group performance during simulator training
8) Decentralizing training facilities
9) Use of other than full scope control room simulators (OTFSCRS) for NPPs that already have full scope simulators

(IAEA-TECDOC-1411)

In the case of the Chernobyl incident, there were human issues and design problems. The human element was the lack of training and familiarization with NPPs. Systems were shut down that were needed for protection. This would not be a situation ever predicted by a designer of the NPP or simulator. I visited the Ukraine about 13 years later and, quite honestly, the people really did not wish to discuss it. It was so devastating.

At Three Mile Island, there were failures in the non-nuclear secondary system followed by a stuck valve in the primary system which allowed a large amount of coolant to escape. Inadequate training and human factors design (a hidden indicator light) that led an operator to manually overriding the cooling system.

A systems approach to training (SAT) at the outset of the training program would have brought this to light, in my opinion.

Chapter X

THE EXPANDING UNIVERSE OF SIMULATION

Simulation is global...

When I moved to Central Florida from Port Washington, Long Island, in 1965, Orlando was a sleepy little town in the orange groves. Agriculture was king. Tourism was here but the attractions were limited: Marineland, Weeki Wachee Springs, Gatorland and Cypress Gardens are the only ones that come to mind. The Naval Training Device Center (NTDC) moved into the old Orlando Air Force Base barracks buildings. Most were without air conditioning. The cadre of folks who made the move were highly motivated and knew that, with simulation, we were on to something. We had no idea how simulation would eventually "go viral". Martin Marrietta, the forefunner of Lockheed Martin was the only company in the area that knew what simulation was all about. The extent of their simulation efforts was a helicopter simulator that used a gigantic model board with television for the visual system. In 1988, the Navy moved its organization to what is known as the Central Florida Research Park, adjacent to the University of

Central Florida. Today, there are several centers of simulation throughout the country, (Figure128), but Central Florida is the epicenter of M&S.

Figure 128—Major Centers of M&S throughout the United States (Okraski)

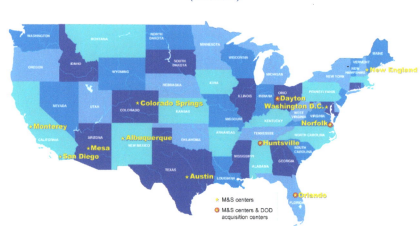

The Florida cluster of companies and government agencies as well as the academic institutions is referred to as the Center of Excellence for Modeling and Simulation. At one time, Senator Bill Nelson referred to Central Florida as a "national asset". When visiting Florida, former president Bill Clinton notes and lauds the large presence of simulation technology industries in this area. Most of the companies are here because of the large acquisition budget of the military, the presence of the University of Central Florida, no state income taxes, proximity to the entertainment attractions and beautiful weather.

The modeling, simulation and training industry has a significant economic impact on the area. (Figure 129). With the downturn of the space program in Florida, M&S has taken the position of lead technology. I like to say that M&S is to the economy as spinach is to the character Popeye. All of the organizations in the area, when clustered, are referred to as "Team Orlando."

Figure 129—MS&T Economic Impact on Florida

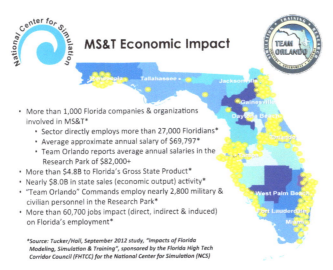

MS&T Economic Impact

- More than 1,000 Florida companies & organizations involved in MS&T*
  - Sector directly employs more than 27,000 Floridians*
  - Average approximate annual salary of $69,797*
  - Team Orlando reports average annual salaries in the Research Park of $82,000+
- More than $4.8B to Florida's Gross State Product*
- Nearly $8.0B in state sales (economic output) activity*
- "Team Orlando" Commands employ nearly 2,800 military & civilian personnel in the Research Park*
- More than 60,700 jobs impact (direct, indirect & induced) on Florida's employment*

Source: Tucker/Hall, September 2012 study, "Impacts of Florida Modeling, Simulation & Training", sponsored by the Florida High Tech Corridor Council (FHTCC) for the National Center for Simulation (NCS)

Team Orlando

How did this cluster of world-class simulation organizations happen to evolve in Central Florida? It began with the Navy. In the United States Navy, the first step in developing and adapting special training methods and devices was initiated by the Bureau of Aeronautics (BuAer). This was effected in April 1941 when Rear Admiral J.H. Towers, then Chief of BuAer, issued a memorandum creating a special device desk in the Engineering Division. The purpose and function of this desk to which then Commander Luis de Florez was assigned, was to supervise experiments and developments of special training devices for primary training, navigation, gunnery training, etc. No funds were set aside for this new unit of BuAer and little progress was made until June, at which time this activity was transferred to the Training Division of BuAer. At that time, funds were set aside for preliminary work.

In 1943, Special Devices became a division of BuAer, and later by executive order was transferred to the Office of Research and Invention, now called the Office of Naval Research. In April 1956, the Special Devices Center became the U.S. Naval Training Device Center, located in Port Washington, New York (on Long Island). The change signified the growing acceptance of training devices by operating forces as well as shore establishments and provided a more descriptive title for the Center's mission. In 1965, the Naval Training Device Center was relocated to Orlando, Florida at the old and mostly vacated Orlando Air Force Base. (Not to be confused with McCoy Air Force Base). A small army unit, the Army Participation Group, joined the navy in the relocation. In 1988, the Center moved to new, modern spaces at the Central Florida Research Park, adjacent to the University of Central Florida. This relocation was not by accident. A few visionaries put their heads together and the "crown jewel" of the proposal was the donation of 40 acres to the Navy to build the facility at was to be known as the Central Florida Research Park. That made the deal very attractive to the Department of Defense. They bought it. Congressman Lou Frey worked up to the 11th hour to move the deal through.

One of the unique features of Central Florida's defense industry is the presence of "Team Orlando". Team Orlando is a loosely-coupled organization made up of the military and other acquisition organizations responsible for the research, acquisition and logistic support of training systems. Team Orlando's success in accomplishing its vital national mission is the result of the unique collaboration between this country's leading military simulation research, development and acquisition commands, along with the integral capabilities, talent, resources and support of private industry, academia and government organizations. (Figure 130).

Figure 130—Team Orlando Members

Team Orlando is a growing community of organizations across defense, government, industry and academia working together to accomplish their respective missions with a common goal of improving human performance through simulation. What is incredible is the synergistic relationships among the services in sharing technologies and simulator "know how". Colocation helps a great deal in technology transfer among the services. Most of the members of Team Orlando are located in the Central Florida Research Park (Figure 131). Joe Wallace is the overall manager of this unique research park and has been the "technological gatekeeper" in selecting park residents that have a common purpose.

Figure 131—Central Florida Research Park (CFRP Photo)

It is truly a "joint" organization-and that is a small "j" and not a capital "J". What this means is that the military services operate jointly in a very synergistic manner without the formality of being a joint organization. Outside of the Pentagon, Orlando is the next "node" where all of the services come together to fulfill their M,S&T missions. (Figure 132).

Figure 132—Simulation and Training Nodes (Okraski)

It is a great break for the taxpayer because of the sharing of the resources and ideas. It is remarkable that the longest active cooperative agreement between services still exists today between the Navy and Army. It goes back to the 1940's. Today or in the future, should any one of the organizations be removed or relocated for some political or other reason, the Orlando model could collapse and the nation would be the loser, in my opinion. Congressman Lou Frey, as mentioned earlier, has been instrumental over the years in keeping Team Orlando intact, fending off attempts in the past to dismember this great resource.(Figure 133). He continues to support M&S in Florida as the counsel for NCS and he sponsors the Lou Frey Institute at UCF which offers stimulating student programs in STEM and M&S.

Figure 133—Congressman Lou Frey and the Author at Crooms
Academy for Information Technology, Sanford, Florida

I have to take a break to tell the story about the Lou Frey "tree". At one time I managed a light industrial simulator fabrication facility for the Navy at Orlando's Herndon Airport. Congressman Frey, who served in the U.S. House of Representatives from 1969-1979, was instrumental in getting us a brand new building built. For the initial ribbon-cutting, several dignitaries were invited along with Congressman Frey. As a part of the celebration, our team planted a young oak tree at the entrance to the new building. There was even a sign at the base of the tree indicating that it was dedicated to the Congressman for his hard work in getting the building funded, etc. Well, time passed and about four years later, word was that he was going to pay us another visit the following day…and he put out the word that he wondered how the tree was doing. The tree died two years before. Not to panic: I dispatched two of our machinists to a nearby clump of woods to cut down a tree, preferably an oak that was about 20 feet tall (I estimated the growth factor). The guys found one and they cut it at its base. I had them stick it in the ground, same place as the original, with guide wires holding it erect. The Congressman came, looked at the tree and was pleased that it was doing so well! We escaped another bullet. About two years ago, I let him know what happened and he could not stop laughing. A real good sport.

While on the topic of buildings, when with the Navy, I escorted many visitors through our modern facility in the Central Florida Research Park which was designed by an architect who studied under Frank Lloyd Wright. The de Florez Building has many features that give it a nautical character: blue exterior water line, waves of earth at the building's bow, traveling reflective waves in the aisle overhead reflectors, etc. Most visitors could not believe it was a government building. More significantly, they were totally impressed with the work that went on there. The imagination of the people in the building is evident when their projects are demonstrated. Also, the range of technologies varies from hard science to the softer disciplines like education and psychology. Simulators for aviation, ships, submarines, tracked vehicles, etc. It didn't matter. Sometimes we heard the extreme. I recall when Florida Governor Lawton Chiles visited he jokingly asked if we could build a simulator for quail hunting. Seems that he was missing too many birds.

The Interservice/Industry Training, Simulation and Education Conference (I/ITSEC)

The largest simulation and training conference in the world is held annually, usually in Orlando, Florida. Thousands of people attend the conference to hear papers delivered, attend panel discussions, take in tutorials and/or walk through the conference exhibition area where the most current technologies are demonstrated. It is truly a magnificent event. The conference is sponsored by the National Training and Simulation Association (NTSA), a branch of the National Defense Industry Association (NDIA). The conference attendance has grown like bacteria over the years. The media gives it a great deal of coverage to include Halldale Publications and Richard Burnett of the Orlando Sentinel.

I attended and presented a paper on Integrated Logistics Support in 1966 at the very first conference of its type (which later became I/ITSEC) and it was held in Orlando at the old Orlando Air Force Base. The first conference was instituted by Dr. Hans Wolff, the NTDC Technical Director at the time. He wanted to initiate discussions between the Navy and Industry. The Army unit, Army Participation Group, was included in the program. It was a very small contingent at that time. (I vividly remember being nervous at the podium as I had not delivered many papers before that time. I felt like Carl Walenda, the high wire artist, with a middle ear infection). The conference attendance was about 50 in number but the discussions were very productive. So, the Navy and Army continued to hold these meetings once a year. In 1975, the first actual I/ITSEC was held in Orlando and I was the Program Chairman. Vince Amico, an icon in our business, was the Conference Chairman. Up until that time all of the presentations were plenary sessions, held in series. With the attendance growing and considering that the interest in specialty areas was also expanding, I decided to initiate parallel sessions. That way, more material could be covered in the same amount of time. Why invite folks to doze off where they have no interest in the subject matter, I rationalized. Most, if not all, conferences of this type are conducted with parallel sessions today. In future conferences, we added tutorials and even technology demonstrations. One year we demonstrated networking live and with virtual simulations. What once was a purely defense conference, I/ITSEC has evolved into a well attended multidimensional conference with medical, education, homeland security and other displays and papers.

Modeling and Simulation
Professional Certification
The CMSP (Certified Modeling & Simulation Professional) certification program was created in 2002 by RADM Fred Lewis and the National Training and Simulation Association (NTSA) to provide the Modeling & Simulation industry with its own professional certification. Like the Professional Engineer (PE) or Project Management Professional (PMP) certifications, the CMSP designation signifies competence, skill and experience in M&S.

Anyone working in the field of Modeling & Simulation—developers, managers, analysts, and users/customers can pursue the CMSP certification. The Modeling & Simulation Professional Certification Commission (M&SPCC)

believes that the CMSP designation will become the industry standard in the years to come, and that in the near future:

1) Industry professionals will be compelled to get the CMSP credential, as it will signify their knowledge and expertise, and further their careers.
2) Companies and organizations will encourage and even require their employees to pursue the CMSP credential—both to ensure that their employees are well-educated and certified, and to demonstrate this fact to their customers.
3) Customers will require companies working for them to have CMSP-certified employees, so they can be assured of competency and professionalism.

References for taking the CMSP exam are provided in Appendix B. For more information, go to: *www.trainingsystems.org*

National Center for Simulation (NCS)

Any discussion about M&S would not be complete without addressing the National Center for Simulation (NCS). NCS is headquartered in Orlando, Florida. We like to say that the National Center for Simulation is located in the national center *OF* simulation-it makes sense, doesn't it? I am proud to say that I am one of the founders of NCS and continue to serve on its Board of Directors. The current Executive Director is retired Lt. General (Air Force) Tom Baptiste.

The National Center for Simulation (NCS) was formed in 1993 as the link among the defense industry, government, and academia on behalf of the entire simulation, training, and modeling community. Its mission is to facilitate networking among its growing local, national, and international membership and potential partners and customers in government, industry, education, and commerce.

NCS was originally created as the Training and Simulation Technology Consortium (TSTC). Under a White House grant program for defense re-investment through technology transfer, TSTC was born. Eight members, four from government and a like number from industry formed, under the management of NASA, to transfer simulation and training technologies from defense to commercial use. Several original players are still very ac-tive. They are Dr. Lee Lacy, Dave Manning, Marcia Bexley, Priscilla Elfrey, Lou Frey and Janet Weisenford. TSTC found that the technology transfer objective was much more difficult than originally perceived. For example, commercial customers want to see a prototype or finished product before they buy. Defense customers issue a request for proposals with a specifica-tion included in the contract. Accounting systems are different. These are just a few of the differences that put TSTC on a road with potholes. Not to be discouraged after a few years of sluggish operation, TSTC assumed a new name, NCS, and adopted a new mission. This turned out to be a win-ner. NCS, starting with 8 members now has over 200 members, much of it due to Tom Baptiste (Figure 133) and his staff. NCS receives continuing support from the State of Florida, the Florida Hi Tech Corridor Council, local city and county governments and other sources. NCS has a very strong Education and Workforce Development Committee that works closely with school districts, teachers, industry and government. NCS, in concert with all catagories named, developed a high school four year curriculum in Model-ing and Simulation. The curriculum is available for teachers and students throughout the nation on the NCS website: www.simulationinformation. com. NCS offers scholarships at UCF and Daytona State College, arranges for teacher and student internships, provides mentors at local schools and provides guest lecturers to schools, companies, social and technical societ-ies, etc. NCS has a close relationship with the University of Central Florida, having worked with the faculty and IST in establishing academic programs in M&S.

Figure 134—Tom Baptiste and Author at Flight Safety, Inc. (Okraski)

As a forum, showcase, advocate and archive for simulation, training, and modeling knowledge and resources, NCS works to strengthen the simulation community's synergy, foster innovation, and tell the story of modeling, simulation, and training to decision makers and the general public. NCS members actively create an environment where collective efforts result in new awareness and applications for military readiness, space exploration, health care, transportation, education, entertainment, and technolog. NCS is a driving force in the world of M&S.

It Takes a Village

We recently organized a team of industry members, educators from Orange and Seminole Counties, Florida, along with members from NAWCTSD, and put together a four year high school curriculum in modeling and simulation. (Figure 134). The program has two tracks: Visual and Software. Several schools are using portions of the curriculum, consistent with state standards. Eventually, we will have full magnet programs in modeling and simulation and fully execute the workforce model shown in our model. (Figure 135).

Figure 135—High School Curriculum in M&S

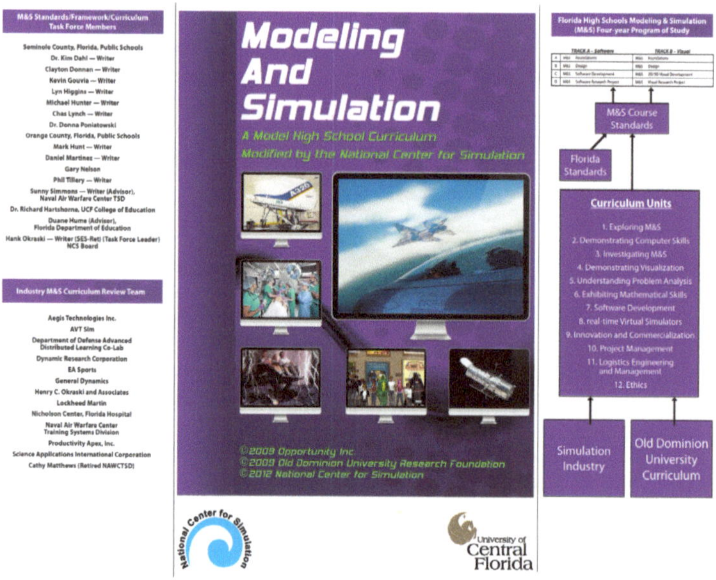

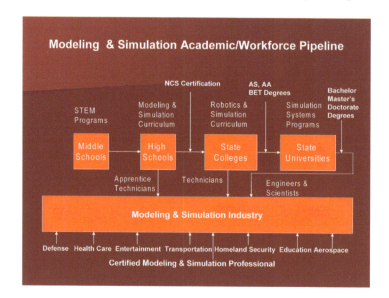

Figure 136—Modeling and Simulation Academic/Workforce Pipeline

The figure shows the broad array of possible M&S careers in industry. It also shows entry points into the M&S job market as apprentice technicians, technicians and engineers and scientists. These are only the technical positions but there are others in fields such as education, psychology, CAD/CAM, accounting, etc. My model purports to resurrect the once successful apprentice program where high school graduates, with certification, can enter the job market and be immediately productive. New hires can go on and work toward degrees with the assistance of their employer. To be competitive in the world market, we have to move in this direction.

The initiatives of the University of Central Florida, Valencia College, Seminole State College and Daytona State College are taking shape to give students many options in Modeling and Simulation. School districts are totally supportive of our initiatives and they work hand-in-hand with us. A good deal of credit must go to Randy Berridge of the Florida High Tech Corridor Council who supports NCS and other programs such as the techPATH Program, operated by Dr. Jeff Bindell and Vicki Morelli. techPATH has educated thousands of teachers in physics, robotics and modeling and simulation.

University of Central Florida (UCF)

UCF, in 2013, celebrated its 50th Anniversary. The university has M&S in its "DNA". From the very beginning, our industry influenced the academic programs offered. When conceived, UCF was going to be the "MIT of the South." This was based on the planned interface with NASA at the Kennedy Space Center, Martin Marietta and the presence of a "simulation" organization-the Naval Training Device Center. The original name of UCF was Florida Technological Unversity (FTU) but later on it was renamed to University of Central Florida to broaden its academic and research programs. We in the government hired junior faculty members who matured and took positions of prominence at the university. Dr. Gary Whitehouse was one who later served as Dean of Engineering and Provost, and there are others. So, they grew up with the simulation industry. No wonder UCF was the first to offer graduate degrees up to the PhD in simulation.They continue to offer an interdisciplinary masters degree and doctorate in simulation. At UCF, the "heavy lifting" in the graduate program is accomplished by Dr. Peter Kincaid. I was fortunate to have participated on advisory boards for setting the curriculum in M&S and I taught there for several years as an adjunct faculty member. (I also taught in that same capacity at Rollins College in Winter Park). Dr. John Hitt, President of the university, has been totally supportive throughout the growth of the university and the simulation industry. Today, there are in excess of 60,000 students at UCF, the second largest university in the nation.

— Chapter XI —

LOOKING AND GIVING BACK

A career in simulation means going to work
every day with a gleam in your eye...

aving exceeded my wildest dreams of a professional career, it is time
to pay back the investment people made in me. As a high school student, I thought I might be an electrician. I graduated from high school
after taking a career and technical education electrical program, under the
watchful eye of my teacher, Henry Guilfoyle. Looking back, our high school,
Utica Free Academy, had a marvelous array of academic and technical programs. One year, for example, a class taking building construction actually
built a house from the ground up. A high school graduate could go to work
at General Electric, for example, and make a career with that company after
taking a program in mechanical design. If that same student decided to go
to college, the program was sufficient in math and science to provide the
needed pre-requisite academic courses. There were options. I decided to en-
roll at Clarkson College and pursue a degree in electrical engineering. I was
following in the footsteps of my long-time friend, Phil Wisniewski. I was
totally prepared for entering an electrical engineering program in college

because I already had first hand experience with motors, generators, circuits, etc in high school. Speaking of Clarkson, in 2008 I was recognized by Clarkson University with their "Golden Knight" award, the highest honor given by the university to alumni, for my career work in simulation and support of the university.

Working my way up the civil service career path, I became a member of the Senior Executive Service (SES). SES is one level above the GS or GM levels-if the reader is familiar with those ratings. The beauty of it all is that I loved what I did and felt proud of the fact that our simulators saved countless numbers of lives. Upon retirement, I turned to private industry-actually I was recruited by a small business where I became a vice president for research. Our team had tremendous success with government SBIRs. I got to work with a very creative professional, Dr. Dave Dryer. His specialty is data visualization. Unfortunately, the company fell on hard times. So, I left the company and did consulting work and emersed myself in volunteer work in Science, Technology, Engineering and Mathematics (STEM) with a focus on modeling and simulation.

Putting all of my accomplishments aside, by and far the most gratifying feelings came from witnessing those who I took under my wing doing well and being successful. I mentored several young people over my career and proud of each and every one of them. I will not list them here but I will give you one example. When Mike Gerrity and I went to the U.S. State Department auditorium for a Systems Effectiveness Conference in the 70s to present the results of the first logistics support verification ever conducted for a training system, I was overjoyed. Mike wrote the paper and was to present it. I was in the audience, quite nervous, but when the audience of what we might call "nerds" today applauded, I had a watermelon sized lump in my throat. There are so many others that I take pride in-and perhaps some "ownership" in their success. They know who they are so I will not try to name them all here. I might leave one out and be angry with myself forever. My mentor, Bob Dreves, (along with Ruby Altman) although gone now, had the same pride in me whenever I did well. When I didn't come up to par, he would give me some special coaching. Vince Amico was always there to provide advice and guidance when I needed it.

The future is with our children and grandchildren. We have to invest the time and energy to make sure they grow up to be moral, productive and responsible citizens. We cannot pass this responsibility to someone else. Now is the time to give back what we owe them and our society.

Chapter XII

THE FUTURE OF SIMULATION

Today is yesterday's future..

I thought it necessary to take out the crystal bowl and make some projections of what the future might bring. In the process of doing so, I rummaged through my files to see what I have said about this in the past. About 20 years ago, I made some statements about the future while I had my finger on the pulse of simulation technology at the Navy. Several of them are still true today. I would like to replay them for you, and you can be the judge of how much validity they have. Here is what I predicted back then:

"The future belongs to simulation. Networking simulators (live-virtual-constructive) will afford new opportunities for team training, especially for joint and combined operations. Embedded training will permit a wide range of training capabilities. The military and civilian worlds will share information "superhighways" and there will be unprecidented commercial "spin-off" and "spin-on" opportunities.

Helmet-mounted display technology, combined with powerful computational systems, will yield dual purpose deployable mission rehearsal and training systems. Artificial intelligence and voice recognition will be used in simulators and deployable trainers. Engineers will be drawn closer to education specialists and psychologists through necessity as reduced budgets will require trade-offs between realism, training effectiveness and cost. Increased emphasis will be placed on computer-based education and training. Look for the commercial entertainment world to take the lead in technology development.

Display technology will give rise to virtual environments. Eyephones, tactile and audio simulation will provide a measure of realism not achievable with current technology. Virtual environments will be created for maintenance, operator and team training. Greater demands will be placed on the (government) engineer to become familiar with new technologies and maintain a "smart buyer" posture. Systems engineering using NDI (Non-developmental Items) will be the mode of operation.

We are in for some interesting and challenging times. The information technology explosion will open new opportunities for us." *(NAWCTSD News)*

Given the above projections, let me move the time pointer to the present as a springboard to the future.

Military Applications

Simulation of all forms is here to stay. Simulation will shape our very lives in the future. I see some very exciting possibilities. First of all, I believe future training simulators will be of designs that extrapolate those of today with increased realism. For example, the trainer developed for the Navy's boot camp, Battle Stations 21, will be improved and utilized heavily in the future. This simulator (Figure 137), represented the combined efforts of the simulator technology community and the entertainment experts who added features to the simulator such as fire, smoke, wave action, etc. just as would be experienced on board ship.

Figure 137—USS Trayer (BST 21)

Vice Admiral Al Harms who was the Chief of Naval Education and Training at the time had a vision of creating an environment that closely emulated shipboard life to acquaint new sailors with the havoc that they might experience in the real world. So, he essentially set a course to bring together the best of navy simulation with the technologies and methods employed in the entertainment world. He thought that if the theme parks could suspend disbelief for its customers, why can't that be done for navy training. The result is the "simulated" ship located at the Recruit Training Command (RTC) at Great Lakes Ill. They christened it the USS Trayer (BST 21).

USS Trayer is a 3/4-scale, 210 feet long mockup of an Arleigh Burke-class destroyer enclosed within a 157,000-square-foot building on board RTC. The trainer uses Hollywood-style special effects to create challenging and realistic training scenarios for recruits. Recruit divisions work through a 12-hour Battle Stations 21 experience as a comprehensive test of the skills and teamwork learned during their eight weeks of basic training at RTC. 12 hours of anything that can happen aboard a ship at sea, from missile attacks that can cause fires to flooding caused by exploding undersea mines. There are wave machines to create a sea state to add to the realism. Heat can be generated to simulate a fire behind a bulkhead, strobe lights simulating an electrical wire dancing around and there is even olfactory and wind simulation to create a full experience The Navy is not forthcoming, I am told, with a full description of all of the things that can be simulated. Obviously, they want to present conditions to surprise the trainees just as they would be at sea.

The requirements for battle training scenarios stemmed from lessons learned from actual events. The terrorist attack on USS Cole (DDG 67) in Yemen in 2000, mine damage to USS Tripoli (LPH 10) in Desert Storm in 1990 and the missile strike on USS Stark (FFG 31) in the Persian Gulf in 1987 have all been incorporated into the scenarios aboard Trayer. The training also simulates conditions similar to historic at-sea mishaps, like the fire on board USS Forrestal (CV 59) in 1967.

Aviation flight simulators will include the finest high fidelity simulators with functional and physical characteristics of the aircraft being simulated. The F-35(JSF) simulator will be the next baseline fighter-type trainer. (Figure 137). Trainers will be deployable and have networking capabilities, operating with a myriad of military assets around the globe.

Figure 137- JSF F-35 Simulator (Phoenix Business Journal)

A new set of trainers will enter the services-the Unmanned Aerial Vehicles (UAV) simulators. Some training will be embedded in the control consoles of the vehicles and there will also be trainers for strategic planning, operations and maintenance. As these trainers join various airspaces, operators will be taught how to operate in space with other aircraft and UAVs.

In the future, there will be high fidelity trainers for all crew personnel. For example, the Boom Operator Trainer (Figure 138) gives the Boom Operator the practice he or she needs to be effective at altitude, filling the fuel tanks of receiving aircraft to continue the mission. The Marine Corps and Army will provide more realistic training for helicopter gunship crews with out-the-door weapons firing. Further, there will be more team training possible through networking Live-Virtual-Constructive simulations.

Figure 138- Boom Operator Trainer

Education Applications

In the field of education, there will be a wide spectrum of applications of simulation such as virtual classrooms for new teachers to experience classroom situations with a variety of students interacting with the prospective teacher. The University of Central Florida has such a system today using avatars and actors. The actors in the current system will be replaced with artificial intelligent players. Simulation will enable the students to participate in historic events such as the signing of the Declaration of Independence. We will be simulating the effects of hurricanes, tornadoes, flooding, etc., giving the student more insight into these phenomena The environment will be studied using simulation: the effects of global warming, for example. Students will be able to experience virtual internships without leaving the classroom, employing avatar technology. The possibilities are almost limitless. Unfortunately, the funding has not been there for education. Hopefully, this will change as the younger generation moves into decision-making positions and who may have a better appreciation of the power of simulation.

Simulation: a Step toward World Peace

There may be a time when wars will become almost obsolete through simulation. Far fetched? I suppose so- but possible. Two countries ready for war might turn to a simulation to predict which country would be victorious in a conflict, given the assets, capabilities, etc. of each. The simulation would be integral to an arbitration conducted to settle their differences. We would have wars without blood. Wouldn't that be wonderful?

Opinions of Futurists

I asked two experts for their opinions on where simulation is heading, Dr. Randy Shumaker, Executive Director of UCF's Institute for Simulation and Training and formerly with the US Office of Naval Research, and Dr. Roger Smith, Chief Technology Officer for Florida Hospital's Nicholson Center, and pioneer in medical simulation. Both of these distinguished gentlemen have considerable experience in simulation. Their views follow.

Future of Simulation
By Roger Smith

Simulation devices and technologies have always been heavily used and promoted by the military, airline, and nuclear industries. Beyond those, the term and practice have often referred to human training methods and scenarios with very little specialized equipment that would be referred to as a simulator.

Those who are part of the simulation industry and who consider themselves simulation scientists and practitioners, usually see the opportunity to apply the technologies beneficially in other industries, though they seldom find the means or customers through which to explore that potential. But we all continue to dream of a time when simulation will become a ubiquitous tool across large numbers of industries that require advanced tools for training, education, and analysis.

Science Fiction authors have presented stories in which the physical world is significantly augmented or mirrored by a simulation-driven world. *Ender's Game, Snowcrash, Halting State, True Names,* and *The Matrix* each envision a future in which the simulated or digital world is equal in importance to the physical world with which it is linked. These encourage simulation practitioners and futurists to search for the growing shadow of change which will bring these visions to reality.

But, perhaps the explosive use of simulation is already happening, though under different names in different industries. For example, for robots to exhibit any type of independent, intelligent behavior, they must have an internal understanding of the world in which they are operating, including some model of how the objects might behave when the robot interacts with them. These digital maps and models are mirror world simulations.

All of the money that we supposedly have "in the bank" is actually a digital record in a simulated world. The money "travels" from one account to another, becoming ownership of a corporate bond or stock in an Australian gold mine without ever becoming a physical object along the way. This digital wealth has a direct impact on our ability to live in the house of our choosing, purchase groceries, or send children to college—all immensely important activities in the "real world." These digital financial records are direct models of the real world.

In this kind of world a simulation professional should find opportunities to apply his or her craft in multiple unique industries if it were just possible to explain the relevance of those simulation skills in a language that is familiar to professionals in other industries.

Custom Technology

Advances in simulation-specific technologies are largely driven by investments in new capabilities by existing customers like the military who already rely heavily on simulation. These advances enable a larger volume or more complex solutions in an area that is already a major user of simulation. Many of these advances fall into interoperability techniques; artificial intelligence models that are uniquely suited to the domain; computer graphics to visualize the physical or information space which is being simulated; and computer hardware for expanding the size, complexity, and interactivity of the virtual world. In most cases, the techniques or products that result from these investments are specialized and not easily applied outside of the military, airline, nuclear, or healthcare field in which they were created.

Imported Technology

Computers and communication networks are essential components of every part of society. This creates a significant financial incentive to make better devices which can be applied and sold to large segments of society, not just a single business domain. When these technologies become available, they are applied to every industry, one of which is simulation. These include computer graphics rendering, internet connectivity, virtual private networks, super computers, cloud computing, cellular networks, and mobile devices. Each of these has brought a new capability to the simulation world and expanded the solutions that are available to current customers, as well as enabling us to attract new customers.

Cloud computing and cellular networks are two recent and extremely complimentary technologies that allow almost any modest computer device to present the results of a significant computer farm. By placing the computing work "in the cloud" and connecting it to any consumer via cellular networks, a cell phone or tablet can be used to perform a task of any computational size. The user no longer needs to carry or even own the power of a super computer in order to tap into it on a moment's notice, use it for a few minutes, and then release the resource for others to use in the same way. The voice-to-text translation feature on smart phones is one very popular example of services that depend on these paired technologies. They would also enable any cell phone to serve as a portal to the largest wargame running on a farm of computer servers. Both 2D maps and 3D virtual scenes can be delivered to the device in real-time, making it a replacement for the dedicated computers of the past. Imported technologies like these will significantly transform the simulation industry as they are adopted.

Exported Simulation

Military simulation has clearly gone through a growth curve, from a time when humans with radios acted as chess pieces on a giant chess board that spanned many square miles of real terrain, to the computerization and automation of these players into semi-intelligent simulated agents in a computer wargame.

Online education, virtual schools, intelligent computer games, remote healthcare, cyber security, and homeland surveillance will create new demands for the simulation of activities which were previously performed by humans, but which can be done adequately by simulated intelligent agents. As Clayton Christensen pointed out in his influential *Innovator's Dilemma*, the disruption of current methods for performing work does not have to be superior in quality to the old method in order to gain a foothold. The new methods simply have to be sufficiently good to satisfy customers who really cannot afford the current method. From that foothold, the technologies will earn the revenue to mature and improve until they are ready to move up the customer chain and satisfy those with the most demanding problems.

This kind of growth will be stimulated by industries and segments of society that are growing in size and social importance. Given the current demographics of the Western world, services for seniors with slightly diminished capabilities, but who are still partially working will potentially be a huge future market for many new products and services. For example, simulations may track the location and health of a senior who is equipped with a slightly enhanced cell phone. New devices may serve as memory reminders or tutors for people who need some guidance through a common household or who are wrestling with a problem that is new to them. A network connected, intelligent wristwatch may detect that your blood pressure at noon on Monday is very different from your blood pressure as measured at the same time over the past month. This might be cause for alarm for people with certain medical conditions.

Social Simulation

Over the last 10-15 years computer games have spread from their original young male demographic to every social, age, economic, and education strata. Almost everyone with the means to own or access a computerized device has used that device as a portal for entertainment and gaming.

This shift was followed by an equally large shift in social networking. These services allow people to maintain large numbers of relationships by sharing news, photos, and activities in near real time. As location-based services are added to these networks, all of the necessary pieces exist to create a dynamic simulation of the activities of everyone within a social group. These simulations could trigger and/or organize spontaneous social events in the real world based on tracking of people in the real world. When ten members of a social group all visit the same theme park at the same time, a simulation of the group's activities could alert them to their coincidental co-location. With this knowledge, the ten friends might rendezvous and spend the day riding roller coasters together. Or they could choose to remain separate, but virtually share their day in photos, chats about the rides, and advice for avoiding long lines. The intersection of social networks and simulations is potentially one of the biggest and broadest opportunities in the immediate future.

Push vs. Pull Demand

Apple products are sold with a "push" in which the designers and marketers create such a great product with a great story around it that the demand comes from the creators and their efforts to impress customers. Alternatively, cellular network bandwidth follows a "pull" method in which a need builds for better service and companies work hard to create networks that can fill that need.

Historically, both "push" and "pull" demand for simulation capabilities has been limited to the military, airline, and nuclear industries. The growth of computer games in the 1990's and 2000's was one instance of a focused "push" of unique simulations to create explosive growth in the electronic entertainment industry. The incorporation of simulations into the brains of robots created a "pull" for simulation technologies which could make these devices smarter about their surrounding environment.

Predicting when and where such a demand will emerge in the future requires understanding the growth of specific demographics, social groups, industries, and government support. Social trends to watch for emerging demand include social networking location-based services, cyber security, aging population, virtual education, homeland surveillance, and new threats to national security. The works of fiction cited earlier are beacons of visionary thinking which continue to stimulate us to think bigger about our industry and to strive to bring some of those big ideas to life.

Future of Simulation
By Randolph Shumaker

As you have seen, simulation has a long history and has been closely linked with meeting real requirements for training. Less obvious is that simulation has been necessarily linked to the capability to do computation in one form or another. In other words implementing a model in an executable form, the simulation. In the very earliest days this computation was accomplished with mechanical machines that effectively calculated using levers, pulleys, cams and sometimes very clever mechanical analogs to the real physical device being simulated. With the advent of electronic analog and then digital computer the complexity of models that could be implemented grew very substantially, harvesting advances in computation, often as one of the first applications. The kinds of displays available have had a similar trajectory and are now at the point where the full audio and visual capacity of humans can be engaged if desired. Effectively we have reached a stage where a human can truly "suspend disbelief" and become immersed in a simulation as if it were reality.

One consequence of the exponential growth in computing and display technology was game technology. Inexpensive personal computers, relatively high performance networks available at home, and advanced techniques for implementing simulations resulted in an explosion of games that in prior eras would have been considered highly advanced simulations. While sometimes considered frivolous use of great technology by the traditional simulation world, the gaming industry has actually provided inexpensive technology back to the "serious" simulation world and opened up entirely new application domains where simulation can play a significant role.

Unlike traditional flight, driving and other vehicle simulations, simulators in the broadest sense don't necessarily have any physical place to sit or stand, these may be nothing to touch or otherwise directly interact with in order to create a sense of being in a real situation. Such virtual world simulations are now regularly used to treat or do research on phobias, posttraumatic stress disorder (PTSD), and traumatic brain injury (TBI). Such virtual environments can be highly immersive where vision, sound and even olfaction are completely synthetic, or they may be a mix of virtual and physical elements. This latter is usually called mixed reality, where the user can manipulate some elements of the simulation, and other elements are entirely synthesized.

Some current applications at the University of Central Florida illustrate how these can be applied in creative ways to some non-traditional applications of simulation, and help provide insight into what might happen next. These projects are not presented as unique, but rather representative of some of the wide-ranging and excellent efforts ongoing within institutions throughout the world. Among recent successes is a project to provide student teachers the equivalent of a flight simulator for classroom management. This effort, TeachLive, allows a student teacher to practice managing a class of avatars representing various personalities likely to be in a classroom. The avatars are highly interactive, a combination of artificial intelligence technology and other techniques to provide realistic body language, facial cues, and above all excellent interactive dialog. The objective of this project was to provide a realistic training environment with minimal equipment costs at the delivery end by using standard components and widely available free internet communication. Initial experience indicates a speedup on the order of a factor of six in acquiring effective skills in classroom management.

Many such applications in healthcare are emerging. Relatively low cost virtual environments are being used to help recovery from PTSD by making it possible to recreate in a controlled manner the kinds of stimuli that would normally evoke symptoms. Studies indicate that this is an effective way to help shorten recovery time for the patient while at the same time reducing costs to deliver the treatment. Game-like and virtual reality simulations have been shown to be effective in treating phobias as well, and game-like virtual environments are rapidly proliferating to assist in training important skills involving interaction with people, particularly in healthcare.

The applications mentioned and many others have been made possible by inexpensive computing, storage, communication and display technology that makes previously unaffordable applications of simulation possible and practical. Gaming technology contributes to the ability to provide the necessary software affordably, and in many ways applications are currently imagination and inertia limited rather than technology limited. That too is changing, with much more exposure of people to emerging applications it is very likely that simulation will experience very significant growth in expected areas, and in some unexpected ones as well.

Obvious simulation applications are teaching and working in science, technology, engineering and math (STEM). Less obvious are ways to provide ways to make STEM topics interesting and available to those who may not currently have an interest. Examples of this are projects that create simulation-based exhibits for museums and science centers that are engaging, visual, interactive, and based on real science. These are often called informal education, code words for "secret learning content" where the objective is both to answer questions, and also raise questions and interest to promote learning. Such applications, directly work-related, educational, or entertainment—perhaps elements of all three will be pervasive and available on everyone's mobile device. In effect, simulation and simulators for many things from education to research will be available wherever and whenever needed.

Final thoughts on the Future

Look to an expansion of the use of autonomous vehicles by the military and in homeland security. The military will continue with the deployment of this technology in the air, land and sea. It has the advantages of stealth operations and safety to our military personnel, while its lethality is unquestionable- ask any terrorist. Unmanned Aerial Vehicles (UAVs) have taken the face off of the enemy. They have been very effective in surveillance and neutralizing terrorist activities throughout the world. If the United States is to increase border security, it must deploy more unmanned vehicles alongthem. To train the "pilots" it will be necessary to provide more simulators to train personnel how to operate and employ the "drones".

The next major conflict between major powers will not be fought with manpower and firepower. It will involve very clever computer experts, sophisticated software and cyber-secure systems. Although such a conflict will not be a "hot" war, it has the potential to create great devastation. Why drop bombs on infrastructures such as bridges, power plants, etc. when they can be rendered inactive or inoperable through computer hacking. Information systems and even conversations on the network will be vulnerable to hacking and interference. Social media will be the "low hanging fruit" of espionage whether it be in a military or civilian domain. Consequently, training must be provided to military and homeland security personnel to level the playing fields and, ideally, stay one step ahead of the "bad guys". Simulation will find a new home in that kind of training whereby various methods for infiltration will be simulated and training provided to recognize hacker activity and also countermeasures can be developed.

Another new frontier for determining training effectiveness of simulation will be in brain wave analysis. Related to this is the discovery that there is a story to be told integral to the training exercise, necessary to capture the participation and imagination of the trainee. The entertainment industry does this quite well in "setting-up" or conditioning the customer before the experience.

Why not do the same for military training?

It is possible to determine the level of awareness a person has under various situations. Therefore, brain waves will be monitored during simulator use and the levels of arousal with varying degrees of simulator fidelity. Current methods of determining training effectiveness of a simulator are costly, not readily accepted by the user and require specialized skill sets to conduct.

Look to more research in studying brain waves in the future. It may be the key to the entire issue of substituting simulators for operational systems. In any event, it will be exciting!!

CHIEF PETTY OFFICER MIKE PHILLIPS' EXPERIENCES AS A TRADEVMAN

Mike entered the Navy at 100 Church Street, New York City, New York on 14 February 1957. He attended Basic Training at Navy Recruit Training Center, Bainbridge, Maryland and Airman Preparatory School, Norman, OK. After AN "P" School he attended 18 weeks of training at TD "A" School at NATTC Memphis, TN.

The following description of his career as a TD is written by Mike Plillips.

AEW Trainers, WV-2M and 2F58, Super Connie Simulators (1957-1960)
Upon my completion of TD A School I received orders to Naval Air Test Center (NATC), Navy Air Station (NAS) Patuxent River, Maryland. I was attached to Airborne Early Warning Training Unit, Atlantic (AEWTRAULANT). AEWTRAULANT provided training to support the Atlantic Fleet Airborne Early Warning (AEW) squadrons flying WV-2 aircraft. The WV-2 aircraft were modified civilian airliner Lockheed Super Constellations (Super Connies).

Supporting the AEW Training were two major training devices, The Super Connie Operational Flight Trainer, Device 2F58 and the WV-2 Tactics

Trainer 15J1C WV-2M. Initially assigned to the 2F58 as an operator/instructor, I have vivid memories of supporting early morning Flight Engineer (FE) training sessions (hops) on the 2F58. Typically, these hops were scheduled for 0400 and 0500 because the hops during normal working hours were reserved for the pilots.

Since pilots were not assigned to these hops, it was the TD's job to fly the plane. The hops were conducted by FE Instructors and they tried to get as many in-flight failures and emergencies they could in the hour assigned for the hop. There would be electrical bus failures on takeoff roll, both before and after takeoff commitment. It was normal to execute multiple takeoffs before the Instructor would allow you to get airborne. All hops were local and you would fly radio aids navigation up and down the Chesapeake Bay until it was time to make your approach back into NAS Pax River. Of course, by then the Instructor FE may have failed three of the four engines on the aircraft and handling the aircraft became quite difficult. With three failed engines, limited hydraulic pressure, and who knows what the electrical failures induced to the aircraft, it became exceedingly difficult to control the aircraft. I can remember heaving the yoke back into my stomach and standing on the rudder pedals trying to keep the Super Connie from falling out of the sky. Woe be it unto you if you ended the hop by crashing the aircraft. The Instructor FEs cherished their hour of instruction and didn't like it when you had to reset the trainer after a crash.

Aviation Physiology Trainers, Low-Pressure Trainer, Device 9A1, Seat Ejection Trainer, Device 6EQ6

For a period of time I was "loaned" to the NAS Link Shack to make up for a shortage of TDs assigned to NAS Pax River. One of my responsibilities was to provide operation and maintenance of the two Aviation Flight Physiology Trainers assigned to NATC Service Test Division. Operating the Low Pressure Chamber, Device 9A1, was not very challenging but, it was interesting to watch the students become confused and stammer when they were required to remove their oxygen masks at altitude.

Perhaps the most dangerous trainer I worked on was the Seat Ejection Trainer. It used black powder cartridges to shoot the seat up 40 foot rails at half the force of an actual aircraft ejection seat, approximately 4-5g. One day I was stationed about 20 feet in front of the trainer observing the indicator panel that showed when the student was properly secured and positioned in the seat. Once I had all green lights on the panel I pushed the pickle switch making the final connection that allowed the seat to fire. However, this time when the face curtain was pulled the seat went up the rails about two feet and settled back down in the cockpit. I couldn't figure out what happened until the entire firing tube and firing mechanism hit the deck about two feet in front of me and knocked a big chuck of concrete out of the apron. The tube and mechanism weighed about 50 pounds and could have killed or seriously injured me.

Shipboard Training System, Talos Missile System Training System, Device 19A1 (1961-1963)

As my TD Class B School approached completion in late 1960 we were all anxious to know where we would be assigned for duty after graduation. Of the seven members of my class five were married with children and myself and my buddy, Floyd Short were either just married, me, or just divorced, Floyd. For some strange reason our class was assigned to fill two shipboard billets. At the time there were no more than a dozen TD shipboard billets and it seemed strange that almost 20% of the billets needed to be filled by our TD B School class. Floyd received orders to the USS Boston CAG-1, and I received orders to USS Little Rock CLG-4.

Irrespective of how I got to the Little Rock, it probably was the best opportunity for me to advance my experience in electronics design and implementation of original ideas in a sophisticated Weapons Systems Training System. My responsibility in this assignment was to operate and repair the Talos Missile System Simulator, Device 19A1. The device simulated the six different radar systems that comprised the Little Rock Weapon System. It was capable of providing six moving radar and IFF targets in three degrees of motion using six miniaturized analog moving target generators that provided simultaneous target position to the six search and fire control radars of the ship. The 19A1 system architecture consisted of six moving target generators in a single unit that also included the operator station. There were three additional cabinets with pull-out drawers that generated the target video for the four search radars and two missile tracking radars. All these simulations were independent of each other and only correlated when displayed on the ship's AN/SPA-34 radar repeaters and the SPG-49 Tracking Radar indicators.

All the radar generators required constant range, bearing, and altitude adjustments to ensure that all targets (36) correlated when the 19A1 was used for crew training. The trainer had a great reputation among the CIC team and the SPG-49 operators for training accuracy. However, the system was a nightmare to calibrate all 36 targets because it required the adjustment of numerous trimpots for each of the simulated radars in the simulator compartment on the main deck Radar 4 space and the visual readout of the trimpot adjustments in CIC and the SPG-49 space. To make the situation even worse was the fact that the trimpots were located on flip up circuit boards in different drawers of the radar target generators. To calibrate one target involved the opening of many drawers and pulling up and securing circuit boards for each radar. Calibration was an ongoing process that took up most of my and a TD "Striker's" time since the simulator used miniature vacuum tubes which drifted continuously.

During a yard update and repair period for the ship the 19A1 was removed from the ship and moved to NTSCLANT Regional Office where a team of electrical engineers and technicians (TDs) were assigned to complete system performance and maintainability improvements. I assisted the engineers in the performance improvement but most of my effort was directed to maintainability improvements. We designed and installed modifications to all the radar simulators and power supply drawer that allowed us to fault isolate problems to functional subsystems of each radar simulator and individual power supplies. This capability was implemented by bringing circuit test point out to the front panel of each simulator drawer and the power unit drawer. These modifications significantly improved trainer maintainability and availability.

Carrier—Aircraft Weapons System Training Systems Experience

Tracker, E-1B (WF-2) CARAEW Aircraft Weapon System Trainer (WST), Device 2D1 (1963-1966)

I was transferred from USS Little Rock CLG-4 to VAW-12 at NAS Norfolk, VA where I was assigned to the maintenance crew of the 2D1. I worked mid-check to work off any malfunctions and prepare the trainer for daily training operation. After I completed AN/APS-80 Radar Maintenance Training I also flew aircrew training hops in the E-1B. This enabled meto calibrate the radar trainer to more closely portray the actual in-flight performance of the AN/APS-80.

Corsair II, A-7A WST, Device 2F84 (1966-1969)

This assignment for me was a major technical breakthrough because the A-7A WST was one of the first training devices designed and implemented with digital electronics technology. The core crew was assigned prior to trainer delivery to NAS Cecil Field, FL. Both the flight and tactics sections of the WST were driven by digital computers. In order to be prepared for factory training, the core crew attended Navy Data Systems Technician "A" School in Great Lakes, IL from February 1966 through May 1966, BRRRRR!

About this time WST fidelity and reliability had improved to the point they were incorporated into training syllabi to a greater degree. It was not unusual for trainers to be scheduled 12-16 hours a day. TD working hours were adjusted to meet the requirements of a three shift operation. The first and second shift was staffed to accommodate supporting training sessions with operators and a small section of TDs that could handle unscheduled maintenance actions. The third shift would be composed of TD maintainers who performed scheduled and unscheduled maintenance actions for all sections of the trainer, flight, tactics, visual, motion, and digital radar landmass simulation.

There was some doubt that TDs could make the transition from analog computer driven simulation to digital computer simulation. A Maintainability Demonstration was conducted after factory WST training that verified that TDs were perfectly capable of maintaining the A-7A WST. In fact, some of the crewmembers designed and implemented a number of modifications to the trainer software that provided timely aircraft system simulation of normal and emergency procedures months before the simulator contractor could have been contracted to provide the simulation. This was critical at the time because we were training attack aircraft pilots prior to their deployment to SE Asia in support of the Viet Nam War. In particular, one modification that I and another TD, Joe N. Wilson, implemented was the simulation of the operation and delivery of the Shrike Missile System including anti-aircraft threats.

Another modification that I provided in a timely manner was the simulation of the engine compressor stall emergency. Compressor stall of the A-7A engine was initially encountered during the first deployment of the Corsair II to Yuma, AZ. Due to the low density of the ambient air in the Yuma area due to the high Outside Air Temperature (OAT), the A-7 was experiencing engine compressor stall during landing approaches. This was especially critical because the compressor stall typically occurred when the aircraft was turning onto final with relatively high angle of bank at 1000' to 1200'. Upon experiencing compressor stall, the engine would unwind to about 37% RPM, with a severe loss of thrust. Without immediate corrective action the aircraft was in danger of an unpleasant "terrain intersect". I designed and installed the compressor stall modification, including the proper resolution of the emergency with corrective action (activating the bleed air valve switch), in a matter of weeks. We were presented Beneficial Suggestion Awards for this modification.

The most memorable event in my Navy career happened at Cecil Field. I was selected for promotion to Chief Petty Officer (CPO) on my first try. There is no other enlisted position in the ranks of the US military that equates to CPO. The Chief is usually the most knowledgeable in his/her rating and has gained respect and confidence of his division. In addition, he/she is now a respected member of an elite group, Chiefs. It makes no difference what your rating is; among fellow chiefs you are an equal. Each Chief knows that they are the backbone of the Navy. It is well known in or out of the Navy if you need a solution to a problem, "Ask the Chief."

Having been "that" Chief has done me very well in my civilian career. I feel that I am still respected and my opinion counts.

Phantom II F-4B WST, Device 2F55 (1969-1970)
I was transferred from NAS Cecil Field, FL to NAS Boca Chica, FL (Key West) to take over the Phantom II F-4B WST, Device 2F55, while I waited for the development and delivery of the F-4J WST, Device 2F88, from Singer Link Co., Binghamton, NY. An oddity of working in trailer-mounted trainer environment is the temperature in the trailers needed to be kept low due to the technology of the vacuum tube and early transistors. These devices were very sensitive to temperature and needed to be kept at lower temperatures to maintain circuit stability. Therefore, it was not unusual for me and my crew to be wearing foul weather jackets and flight jackets when leaving our office and shop for the F-4B trailers behind the building. It was not unusual for the temperature in the trailers to be 60 degrees F while the outside temperature was better than 90 degrees F. They got lots of looks from sailors who were not aware of our situation.

Phantom II F-4J WST, Device 2F88 (1971-1974)
While waiting at NAS Boca Chica for the delivery of the F-4J WST, the Navy pulled a swap on the F-4 community. They decided to move VF-101, the F-4 Replacement Aircrew Group Squadron, to NAS Oceana, VA. That meant that I would need to move the Device 2F55 to Oceana, along with my

growing family to Virginia. I was informed by the FASOTRAGRULANT Oceana there was no office or shop space available at Oceana for the 2F55 crew so I had to improvise. At NAS Key West there was an empty prefabricated building close to the Device 2F55 site. So one Saturday morning just before the Device 2F55 trailers were to be shipped to NAS Oceana the crew disassembled and loaded the prefab building into the Device 2F55 trailers, including a water cooler. The building was reassembled and mounted on a wooden platform between the Device 2F55 trailers at NAS Oceana. The building still had a NAS Key West building number painted on it.

Myself and a cadre of his First Class Petty Officers traveled to Singer Link to take an eleven week training course on the Singer Link Digital Simulator Computational System, the GP-4B Computer. Since the F-4J WST was still being built we used the Link L-1011 and 747 simulators for lab sessions.

The Device 2F88 at Oceana suffered an ignominious situation at the beginning of its life at NAS Oceana. Because the trainer delivery destination was changed from NAS Key West to NAS Oceana mid way through trainer development the Military Construction funds associated with the trainer building needed to change from Key West to Oceana. Consequently the trainer building was not available at trainer delivery. Temporary space was indentified in the Ground System Equipment (GSE) hanger on the flight line. The trainer was self-contained with aluminum panels creating an enclosure building set up on the hanger floor. The trainer was located in the GSE hanger for a year, during which we experienced an infestation of pigeons living on the roof and GSE diesel exhaust sucked into the air-conditioning system. The diesel exhaust was sucked up into all the equipment racks by their cooling fans. Of course we took action to preclude further damage by both factors by draping heavy gauge plastic sheets over the enclosure to prevent further pigeon poop from running down the walls permanently staining them. All GSE were required to turn-up for repair and checkout far enough away from the enclosure to minimize induction of exhaust fumes.

As if this was not enough humility for the F-4J WST, we encountered another major problem when it came to moving the trainer into the new facility. Link facility engineers had approved the construction drawings basing their decision on the dimensions of the building, including the trainer access door. The engineers were correct that all the dimensions agreed with the trainer component requirements. However, they had not consulted with the installation technicians who would have told them that the access door was not correctly located to allow the enclosure installation equipment (forklifts) to maneuver with the wall and ceiling panels in the only order that would allow the enclosure to be erected in the new building. The solution was to tear out a wall in the correct location to allow erecting the enclosure. Now problem two arises. The major components of the trainer (cockpit, motion base...) were transferred from the GSE hanger by a huge forklift. The ground between the road and the new access to the building was not stable enough to allow the heavy fork lift and the major components to transverse the 75 yards without the forklift bogging down in the soft Southern Virginia earth. The solution was to build a temporary road to the trainer building access hole. Unfortunately, all this happened just off the main road on NAS Oceana visible to everyone coming and going on the Base.

The Device 2F88 was a major improvement over the Device 2F55 in many ways. From a performance point of view the two trainers were like night and day. The Device 2F55 flight dynamics were abysmal. One of the fighter pilots most important maneuvers was the "re-attack." After an initial unsuccessful pass at a bogey, the pilot initiates a "4-G" turn at a high angle of bank to come around on the bogey's tail. The Device 2F55 flight dynamics could not hold the "4-G" high angle of bank and would unrealistically fall off. Phantom II pilots dreaded Device 2F55 trainer sessions because they felt they were getting negative training.

However, the Device 2F88 flight dynamics were so good that aircrews would voluntarily fly training sessions. Some aircrews even used the Device 2F88 as a research tool. They would have the TD operator set a Soviet Bear bogey at 100 miles and simulate a scramble from the carrier flight deck to see how far away from the "Fleet" they could intercept the Bear and neutralize it.

Oceana was the home for F-4 and A-6 aircraft squadrons. The trainers for these aircraft were collocated at the FASOTRAGRULANT Detachment. At the time of the state-of-the-art Device 2F88 installation and delivery the A-6 community was still using a trailer mounted WST, Device 2F66. Even though it employed a digital computer for some of its simulation the Device Device 2F66 was antiquated compared to the Device 2F88 performance and instructional capabilities. Many of the A-6 aircrews would visit the Device 2F88 to try to get a flight. They began to voice dissatisfaction with their training assets that eventually led to a new A-6E WST, Device 2F114.

Hawkeye Airborne Early Warning Aircraft. E2C tactics Trainer, Device 15F8

I was selected for the Associate Degree Completion Program in 1975 and attended Tidewater Community College in Virginia Beach, for 18 months. After my graduation I was assigned to FASOTRAGRULANT HQ at NAS Norfolk, VA. Initially, I was assigned as Division Officer of the Aviation Training Aids Division (ATAD). The Division included the E-2C and SH-2F Training Devices and the Film Library and Training Devices and Aides Issue group. Because of technical and "political" issues with the E-2C TT, Device 15F8, I spent the majority of my time working with and directing the trainer crew. The Device 15F8 was originally scheduled to be installed at NAS Glynco, Brunswick, GA, however, NAS Glynco was closed prior to delivery. The E-2C AEW Crew training was moved to NAS Norfolk, VA. After some building modifications, the Device15F8 was installed in a building originally constructed to house a WWII Celestial Navigation Trainer. The air conditioning installed to service was not capable of supporting the heat load of the E-2C TT with its seven networked computers and many heat generating displays. The trainer was enclosed in an aluminum panel enclosure within the building and many times actual moisture would be running on the walls. After much finger pointing between Navy Facilities Department, the Device15F8 contractor (Grumman) and us it was decided that the air conditioning needed to be increased to handle the heat load. Unfortunately, the Digital Radar Landmass Simulation (DRLMS) contractor (Link, Sunnyvale) said that their custom-wire-wrapped circuit cards had suffered mold contamination during the high moisture period and blamed all their DRLMS problems on the mold. There were numerous microscopic inspections of DRLMS boards that lead to a funded replacement of many circuit boards. This and the improved a/c seemed to minimize further DRLMS problems.

A trainer crew initiated action significantly improved trainer availability. There were many actual aircraft units used in the trainer. These were all not certified for flight units. Consequently when they failed and were inducted into the Navy repair system they were marked for "expeditious repair" which meant the units were fault analyzed, repaired, and returned to the trainer. This process could take days because aircraft units would get "head of the line" treatment and bump the trainer units to the rear of the line. Without the failed unit repaired, trainer availability suffered severely. The solution to this problem was to convert all trainer aircraft units to air-certified. After many high level meetings with those who had skin in the game, VAW Wing, NAVAIR, AIRLANT, NTEC and us, FASOTRGRULANT, an agreement was made to the process and funding to accomplish this solution. The conversion happened over a number of months of shipping the units to Grumman Bethpage, LI, NY. After all units were converted the trainer crew would turn the failed unit into the supply system and draw an operational unit. A significant trainer availability improvement was realized.

I finished my Naval career as a Senior Chief TRADEVMAN at FASO-TRAGRULANT HQ. I was the Aviation Trainer Maintenance Manager for over 40 training systems at the FASO Dets stretching from NAS Brunswick, ME to NAS Key WEST, FL. My responsibility was to budget for and allocate maintenance funding for the FASO Dets and oversee maintenance problems and provide innovative solutions. I also was the Trainer Support

Representative on the Fighter/Attack/AEW Trainer Procurement Fleet Project Teams. I retired in February 1980 after 23 wonderful years in the Navy's best rating, **TRADEVMAN**.

CERTIFIED MODELING AND SIMULATION PROFESSIONAL (CMSP)

Examination Question Sources

All sources used for questions as of January 2013 are included; list subject to change and expansion as new questions are added.

E. Aleisa, M. Al-Ahmad, and A. M. Taha, "Design and management of a sewage pit through discrete-event simulation", *SIMULATION: Transactions of the Society for Modeling and Simulation International*, Vol. 87, No. 11, November 2011, pp. 989-1001.

R. B. Allen, "Mental Models and User Models", in M. G. Helander, T. K. Landauer, and P. V.

Prabhu (Editors), *Handbook of Human-Computer Interaction, Second Edition*, Elsevier, Amsterdam Netherlands, 1997, pp. 49-63.

J. D. Anderson, *Modern Compressible Flow with Historical Perspective, Second Edition*, McGraw-Hill, Boston MA, 1990.

C. Anderson, P. Konoske, J. Davis, and R. Mitchell, "Determining How Functional Characteristics of a Dedicated Casualty Evacuation Aircraft Affect Patient Movement and Outcomes", *Journal of Defense Modeling and Simulation: Applications, Methodology, Technology*, Vol. 7, No. 3, July 2010, pp. 167-177.

D. R. Anderson, D. J. Sweeney, and T. A. Williams, *Statistics for Business and Economics, Tenth Edition*, Thomson South-Western, Mason OH, 2008.

D. R. Anderson, D.J. Sweeney, and T. A. Williams, *Essentials of Statistics for Business and Economics, Tenth Edition*, Mason OH, Thomson South-Western, 2008.

A. Anglani, A. Grieco, M. Pacella and T. Tolio, "Object-Oriented Modeling and Simulation of Flexible Manufacturing Systems: a Rule-Based Procedure", *Simulation Modelling Practice and Theory*, Vol. 10, pp. 209–234, 2002.

F. Babich, M. Comisso, A. Dorni, F. Barisi, M. Driusso, and A. Manià, "Discrete-time simulation of smart antenna systems in Network Simlator-2 using MAT-LAB and Octave", *SIMULATION: Transactions of the Society for Modeling and Simulation International*, Vol. 87, No. 11, November 2011, pp. 932-946.

O. Balci, "Verification, Validation, and Testing", in J. Banks, *Handbook of Simulation: Principles, Methodology, Advances, Applications, and Practice*, John Wiley & Sons, New York NY, 1998, pp. 335–393.

C. M. Banks, "Introduction to Modeling and Simulation", in J. A. Sokolowski and C. A. Banks, *Modeling and Simulation Fundamentals: Theoretical Underpinnings and Practical Domains*, John Wiley & Sons, Hoboken NJ, 2010, pp. 1-24.

C. M. Banks, "What Is Modeling and Simulation?", in J. A. Sokolowski and C. M Banks, (Editors), *Principles of Modeling and Simulation: A Multidisciplinary Approach*, John Wiley & Sons, Hoboken NJ, 2009, pp. 3-24.

J. Banks, "Principles of Simulation", in J. Banks (Editor), *Handbook of Simulation: Principles, Methodology, Advances, Applications, and Practice*, John Wiley & Sons, New York NY, 1998, pp. 3-30.2

J. Banks, J. S. Carson, B. L. Nelson, and D. M. Nicol, *Discrete-Event System Simulation, Fifth Edition*, Prentice Hall, Upper Saddle River NJ, 2010.

J. Banks, J. S. Carson, B. L. Nelson and D. M. Nicol, *Discrete-Event System Simulation, Fifth Edition*, Prentice Hall, 2010.

M. Bolas and I. McDowall, "Visual Displays: Head-Mounted Displays", in D. Nicholson, D.

Schmorrow, and J. Cohn (Editors), *The PSI Handbook of Virtual Environments for Training and Education, Volume 2: VE Components and Training Technologies*, Praeger Security International, Westport CT, 2009, pp. 48-62.

C. Bowers, J. Vogel-Walcutt, and J. Cannon-Bowers, "The Role of Individual Differences in Virtual Environment Based Training", in D. Schmorrow, J. Cohn, D. Nicholson (Editors), *The PSI Handbook of Virtual Environments for Training and Education, Volume 1: Learning, Requirements, and Metrics*, Praeger Security International, Westport CT, 2009, pp. 31-49.

K. K. Boyer and R. Verm, *Operations & Supply Chain Management for the 21st Century*, South-Western, Mason OH, 2010.

G. E. P. Box, J. S. Hunter, and W. G. Hunter, *Statistics for Experimenters: Design, Innovation and Discovery*, John Wiley & Sons, Hoboken NJ, 2005.

E. F. Brigham and L. C. Gapenski, *Financial Management Theory and Practice, Sixth Edition*, The Dryden Press, Orlando FL, 1991.

D. E. Brownlee, "The Origin and Early Evolution of the Earth", in M. C. Jacobson, R. J. Charlson, H. Rodhe, and G. H. Orians (Editors), *Earth System Science*, Elsevier, London, 2000.

P. E. Castro, E. Antonsson, D. T. Clements, J. E. Coolahan, Y. Ho, M. A. Horter, P. K. Khosla, J. Lee, J. L. Mitchiner, M. D. Petty, S. Starr, C. L. Wu, and B. P. Zeigler, *Modeling and Simulation in Manufacturing and Defense Systems Acquisition: Pathways to Success*, National Academy Press, Washington DC, 2002.

P. D. Cha, J. J. Rosenberg and C. L. Dym, *Fundamentals of Modeling and Analyzing Engineering Systems*, Cambridge University Press, 2000.

R. J. Charlson and S. Emerson, in M.C. Jacobson, R.J. Charlson, H. Rodhe, and G. H. Orians (Editors) *Earth System Science*, Elsevier, London, 2000.

R. J. Charlson, "The Atmosphere", in M.C. Jacobson, R.J. Charlson, H. Rodhe, and G. H. Orians (Editors) *Earth System Science*, Elsevier, London, 2000.

D. G. Chernoguz, "The system dynamics of Brooks' Law in term production", *SIMULATION: Transactions of the Society for Modeling and Simulation International*, Vol. 87, No. 11, November 2011, pp. 947-975.

J. K. Cochran, G. T. Mackulak, and P. A. Savory, "Simulation Project Characteristics in Industrial Settings", *Interfaces*, Vol. 25, No. 4, July/August 1995, pp. 104-113.

J. Cohn, "Building Virtual Environment Training Systems for Success", in D. Schmorrow, J. Cohn, D. Nicholson (Editors), *The PSI Handbook of Virtual Environments for Training and Education, Volume 1: Learning, Requirements, and Metrics*, Praeger Security International, Westport CT, 2009, pp. 193-207.3

W. N. Colley, "Modeling Continuous Systems", in J. A. Sokolowski and C. M. Banks (Editors), *Modeling and Simulation: Theoretical Underpinnings and Practical Domains*, Wiley & Sons, Hoboken, 2010, pp. 99-129.

R. D. Cook, *Finite Element Modeling for Stress Analysis*, John Wiley & Sons, Hoboken NJ, 1995.

S. R. Covey, *Principle-Centered Leadership*, Simon & Schuster, New York NY, 1991.

M.E. Davis, *Numerical Methods & Modeling for Chemical Engineers*, John Wiley & Sons, Hoboken NJ, 1984.

P. K. Davis and J. H. Bigelow, *Experiments in Multiresolution Modeling*, RAND National Defense Research Institute, Santa Monica CA, 1988.

B. Diamond, et al., *Extend v6 Developer's Reference*, Imagine That, Inc., San Jose CA, 2002.

J. W. Diem, J. S. Smith, and L. A. Butler, "United States Army Operational Test Command (USAOTC) Integrated Technologies Evolving to Meet New Challenges—A Study in Cross Command Collaboration", *Proceedings of the International Test and Evaluation Association Annual Symposium*, Baltimore MD, September 29-October 1 2009.

R. C. Dorf and R. H. Bishop, *Modern Control Systems, Tenth Edition*, Pearson Prentice Hall, Upper Saddle River NJ, 2005.

DSEEP Product Development Group, *IEEE P1730/Dv3.0 Draft Recommended Practice for Distributed Simulation Engineering and Execution Process (DSEEP)*, The Institute for Electrical and Electronics Engineers, New York NY, 2008.

C. L. Dym, *Principles of Mathematical Modeling, Second Edition*, Elsevier Academic Press, Burlington MA, 2004.

R. E. Eberts, "Computer-Based Instruction", in M. G. Helander, T. K. Landauer, and P. V.

Prabhu (Editors), *Handbook of Human-Computer Interaction, Second Edition*, Elsevier, Amsterdam, Netherlands, 1997, pp. 825-846.

M. Eisenberg, "End-User Programming", in M. G. Helander, T. K. Landauer, and P. V. Prabhu (Editors), *Handbook of Human-Computer Interaction, Second Edition*, Elsevier, Amsterdam Netherlands, 1997, pp. 1127-1146.

C. E. Farrell, S. F. Sichi, and K. N. Vu, "Spacecraft Design and Development", in L. B. Rainey (Editor), *Space Modeling and Simulation: Roles and Applications Throughout the System Life Cycle*, The Aerospace Press, El Segundo CA, 2004, pp. 225-287.

G. S. Fishman, "Discrete-Event Simulation: Modeling, Programming, and Analysis", Springer-Verlag, New York NY, 2001.

P. A. Fishwick and H. Park, "Queue Modeling and Simulation", in J. A. Sokolowski and C. M. Banks (Editors), *Principles of Modeling and Simulation: A Multidisciplinary Approach*, John Wiley & Sons, Hoboken NJ, 2009, pp. 71-90.

P. A. Fishwick, *Simulation Model Design and Execution*, Prentice Hall, Upper Saddle NJ, 1995.

M. D. Fontaine, D. P. Cook, C. D. Combs, J. A. Sokolowski, and C. M. Banks, "Modeling and Simulation: Real-World Examples", in J. A. Sokolowski and C. M. Banks (Editors), *Principles of Modeling and Simulation: A Multidisciplinary Approach*, Hoboken, Wiley & Sons, 2009, pp.181–198

T. Forester and P. Morrison, *Computer Ethics: Cautionary Tales and Ethical Dilemmas in Computing, Second Edition*, The MIT Press, Cambridge MA, 1994.

J. Fowlkes, K. Neville, R. Nayeem, and S. E. Dean, "Training Advanced Skills in Simulation Based Training", in D. Schmorrow, J. Cohn, D. Nicholson (Editors), *The PSI Handbook of Virtual Environments for Training and Education, Volume 1: Learning, Requirements, and Metrics*, Praeger Security International, Westport CT, 2009, pp. 251-265.

D. Frost and G. F. Knapp, "The Joint National Training Capability Strategic Framework", *Proceedings of the 2005 Interservice/Industry Training, Simulation and Education Conference*, Orlando FL, November 28-December 1 2005.

R. M. Fujimoto, "Parallel and Distributed Simulation", in J. Banks (Editor), *Handbook of Simulation: Principles, Methodology, Advances, Applications, and Practice*, John Wiley & Sons, New York NY, 1998, pp. 429-464.

D. L. Giadrosich, *Operations Research Analysis in Test and Evaluation*, American Institute of Aeronautics and Astronautics, Washington DC, 1995.

A. Gordon, "Story Based Learning Environments", in D. Nicholson, D. Schmorrow, and J. Cohn (Editors), *The PSI Handbook of Virtual Environments for Training and Education, Volume 2: VE Components and Training Technologies*, Praeger Security International, Westport CT, 2009, pp. 378-392.

J. D. Gould, S. J. Boies, and J. Ukelson, "How to Design Usable Systems", in M. Helander, T. K. Landauer, and P. V. Prabhu (Editors), *Handbook of Human-Computer Interaction, Second Edition*, Elsevier, 1997, pp. 231-254.

A. Greasley, *Enabling a Simulation Capability in the Organisation*, Springer-Verlag, London UK, 2008.

P. S. Greenlaw, L. W. Herron, and R. H. Rawdon, *Business Simulation in Industrial and University Education*, Prentice Hall, Englewood Cliffs NJ, 1962.

D. T. Greenwood, *Classical Dynamics*, Prentice Hall, Upper Saddle River NJ, 1977.

K. K. Gupta and J. L. Meek, *Finite Element Multidisciplinary Analysis*, AIAA, Reston VA, 2000.

D. Halliday, R. Resnick, and J. Walker, *Fundamentals of Physics, Sixth Edition*, John Wiley & Sons, New York NY, 2001.

P. E. Hart, N. J. Nilsson, and B. Raphael, "A Formal Basis for the Heuristic Determination of Minimum Cost Paths", *IEEE Transactions on Systems Science and Cybernetics SSC4*, Vol. 4, No. 2, 1968, pp. 100–107.

A. Hartmann and H. Schwetman, "Discrete-Event Simulation of Computer and Communication Systems", in J. Banks (Editor), *Handbook of Simulation: Principles, Methodology, Advances, Applications, and Practice*, John Wiley & Sons, New York NY, 1998, pp. 659-676.

S. Henderson and S. Feiner, "Mixed and Augmented Reality for Training", in D. Nicholson, D. Schmorrow, and J. Cohn (Editors), *The PSI Handbook of Virtual Environments for Training and Education, Volume 2: VE Components and Training Technologies*, Praeger Security International, Westport CT, 2009, pp. 135-156.

F. S. Hiller, G. J. Lieberman, *Introduction to Operations Research, Sixth Edition*, Mc-Graw-Hill, New York NY, 2005.

C. Hirsch, *Numerical Computation of Internal and External Flows, Volume 1: Fundamentals of Computational Fluid Dynamics, Second Edition*, Elsevier, Amsterdam, 2007.

J. D. Hollan, B. Bederson, and J. I. Helfman, "Information Visualization", in M. G. Helander, T. K. Landauer, and P. V. Prabhu (Editors), *Handbook of Human-Computer Interaction, Second Edition*, Elsevier, Amsterdam Netherlands, 1997, pp. 33-48.

R. Illner, C. S. Bohun, S. McCollum and T. van Roode, *Mathematical Modelling: A Case Studies Approach*, American Mathematical Society, Providence RI, 2005.

The Institute for Electrical and Electronics Engineers, *IEEE 1516.3 Recommended Practice for High Level Architecture (HLA) Federation Development and Execution Process (FEDEP)*, New York NY, 2003.

The Institute for Electrical and Electronics Engineers, *IEEE Standard Glossary of Modeling and Simulation Terminology*, New York NY, 1989.

J. Jackman, M. Richardson, P.Yuen, D. James, B. Butters, R. Walmsley, and N. Millwood, "The Effect of Pre-emptive Flare Deployment on First Generation Man-Portable Air-Defence (MANPAD) Systems", *Journal of Defense Modeling & Simulation*, Vol. 7 No. 3, July 2010, pp. 181–189.

M. C. Jacobson, R. J. Charlson, and H. Rodhe, "Introduction", in M. C. Jacobson, R. J. Charlson, H. Rodhe, G. H. Orians (Editors), *Earth System Science*, London Elsevier, 2000.

R. de Jonckheere and B. Preiss, "Using Simulations", in L. B. Rainey (Editor), *Space Modeling and Simulation: Roles and Applications Throughout the System Life Cycle*, The Aerospace Press, El Segundo CA, 2004, pp. 95-128.

W. Kaplan, *Advanced Calculus, Fourth Edition*, Addison-Wesley, Redwood City CA, 1991.

B. Knerr and S. Goldberg, "Dismounted Combatant Simulation Training Systems", in J. Cohn, D. Nicholson, and D. Schmorrow, (Editors) *The PSI Handbook of Virtual Environments for Training and Education, Volume 3: Integrating Systems, Training Evaluations, and Future Operations*, Praeger Security International, Westport CT, 2009, pp. 232-242.

A. Koochaki and S. M. Kouhsari, "Simulation of simultaneous unbalances in power system transient stability analysis", *SIMULATION: Transactions of the Society for Modeling and Simulation International*, Vol. 87, No. 11, November 2011, pp. 976-988.

E. Kreyszig, *Advanced Engineering Mathematics, Seventh Edition*, John Wiley & Sons, Hoboken NJ, 1993.

E. Kreyszig, *Advanced Engineering Mathematics, Ninth Edition*, John Wiley & Sons, Hoboken NJ, 2006.

O. Labarthe, B. Espinasse, A. Ferrarini and B. Montreuil, "Toward a Methodological Framework for Agent-Based Modelling and Simulation of Supply Chains in a Mass Customization Context", *Simulation Modelling Practice and Theory*, Vol. 15, Iss. 2, February 2007, pp. 113–136.

H. C. Lane and L. Johnson, "Intelligent Tutoring and Pedagogical Experience Manipulation in Virtual Learning Environments", in D. Nicholson, D. Schmorrow, and J. Cohn (Editors), *The PSI Handbook of Virtual Environments for Training and Education, Volume 2: VE Components and Training Technologies*, Praeger Security International, Westport CT, 2009, pp. 393-406.

A. M. Law and W. D. Kelton, *Simulation Modeling and Analysis, Second Edition*, Mc-Graw-Hill, New York NY, 1991.

K. Lin, "Dead Reckoning and Distributed Interactive Simulation", in T. L. Clarke (Editor), *Distributed Interactive Simulation Systems for Simulation and Training in the Aerospace Environment*, SPIE Critical Reviews of Optical Science and Technology, Vol. CR58, SPIE Press, 1995.

L. M. Leemis and S. K. Park, *Discrete-Event Simulation: A First Course*, Pearson Prentice Hall, Upper Saddle River NJ, 2006.

R. Lockhart and C. Ferguson, "Joint Mission Environment Test Capability", *ITEA Journal*, Vol. 29, 2008, 160–166.

R. B. Loftin, "The Future of Simulation", in J. A. Sokolowski and C. M Banks, (Editors), *Principles of Modeling and Simulation: A Multidisciplinary Approach*, John Wiley & Sons, Hoboken NJ, 2009, pp. 247-255.

G. L. Lohse, "Models of Graphical Perception", in M. G. Helander, T. K. Landauer, and P. V. Prabhu (Editors), *Handbook of Human-Computer Interaction, Second Edition*, Elsevier, Amsterdam, Netherlands, 1997, pp. 107-135.

L. E. Malvern, *Introduction to the Mechanics of a Continuous Medium*, Prentice Hall, Englewood Cliffs NJ, 1969.

M. S. Manivannan, "Simulation of Transportation and Logistics Systems", in J. Banks (Editor) *Handbook of Simulation: Principles, Methodology, Advances, Applications, and Practice*, Wiley & Sons, New York NY, 1998, pp. 571–604.

G. Marsaglia, "Random Number Generators", *Journal of Modern Applied Statistical Methods*, Volume 2, 2003, pp. 2–13.

M. Matsumoto and T. Nishimura. "Mersenne Twister: A 623-Dimensionally Equi-distributed Uniform Pseudo-Random Number Generator", *ACM Transactions on Modeling and Computer Simulation*, Volume 2, pp. 2–13, 1998.

T. Mayfield and D. Boehm-Davis, "Applied Methods for Requirements Engineering", in D. Schmorrow, J. Cohn, and D. Nicholson, D. (Editors) *The PSI Handbook of Virtual Environments for Training and Education, Volume 1: Learning, Requirements, and Metrics*, Praeger Security International, Westport CT, 2009, pp. 131-147.

P. McDowell, "Games and Gaming Technology for Training", in D. Nicholson, D. Schmorrow, and J. Cohn (Editors), *The PSI Handbook of Virtual Environments for Training and Education, Volume 2: VE Components and Training Technologies*, Praeger Security International, Westport CT, 2009, pp. 205-218.

P. McDowell, M. Guerrero, D. McCue, and B. Hollister, "Rendering and Computing Requirements", in D. Nicholson, D. Schmorrow, and J. Cohn (Editors), *The PSI Handbook of Virtual Environments for Training and Education, Volume 2: VE Components and Training Technologies*, Praeger Security International, Westport CT, 2009, pp. 173-189.

F. D. McKenzie, "System Modeling: Analysis and Operations Research", in J. A. Sokolowski and C. M. Banks (Editors), *Modeling and Simulation Fundamentals:*

Theoretical Underpinnings and Practical Domains, John Wiley & Sons, Hoboken NJ, 2010, pp. 147-180.

P. McMurry, L. Fulton, M. Brooks, and J. Rogers, "Optimizing Army Medical Department Officer Accessions", *Journal of Defense Modeling and Simulation: Applications, Methodology, Technology,* Vol. 7, No. 3, 2010, pp. 133-143.

C. J. Metvier, C. Gaughan, S. Callant, L. E. McGlynn, J. McDonnell, G. Smith, and K. Snively, "Modeling Architecture for Technology Research and Experimentation (MATREX): M&S Tools and Resources Enabling Critical Analyses", *Modeling and Simulation Information Analysis Center,* 2009, pp. 4–10.

R. R. Mielke, "Statistical Concepts for Discrete Event Simulation", in J. S. Sokolowski and C. M. Banks (Editors), *Modeling and Simulation: Theoretical Underpinnings and Practical Domains,* John Wiley & Sons, Hoboken NJ, 2010, pp. 25–56.

R. Mitchell, J. Parker, M. Galarneau, and P. Konoske, "Statistical Modeling of Combat Mortality Events by Using Subject Matter Expert Opinions and Operation Iraqi Freedom Empirical Results from the Navy-Marine Corps Combat Trauma Registry", *Journal of Defense Modeling and Simulation: Applications, Methodology, Technology,* Vol. 7, No. 3, 2010, pp. 145-155.

M. Mujica, M. A. Piera and M. Narciso, "Revisiting State Space Exploration of Timed Coloured Petri Net Models to Optimize Manufacturing System's Performance", *Simulation Modelling Practice and Theory,* Vol. 18, Iss. 9, October 2010 pp. 1225–1241.

M A. Muqsith, H. S. Sarjoughian, D. Huang, and S. S. Yao, "Simulating adaptive serviceoriented software systems", *SIMULATION: Transactions of the Society for Modeling and Simulation International,* Vol. 87, No. 11, November 2011, pp. 915-931.

K. J. Musselman, "Guidelines for Success", in J. Banks (Editor), *Handbook of Simulation: Principles, Methodology, Advances, Applications, and Practice,* John Wiley & Sons, New York NY, 1998, pp. 721–743.

M. Narciso, M. A. Piera, and A. Guasch, "A Methodology for Solving Logistic Optimization Problems through Simulation", *SIMULATION: Transactions of The Society for Modeling and Simulation International,* Vol. 86, Nos. 5-6, May-June 2010, pp. 369–389.

E. Ndefo, J. Encalada, W. Hallman, W. Wang, and P. Abbot, "Launch Vehicle Design and Development", in L. B. Rainey (Editor), *Space Modeling and Simulation: Roles and Applications Throughout the System Life Cycle,* The Aerospace Press, El Segundo CA, 2004, pp. 289-368.

R. S. Nickerson and T. K. Landauer, "Human-Computer Interaction: Background and Issues", in M. G. Helander, T. K. Landauer, and P. V. Prabhu (Editors), *Handbook of Human-Computer Interaction, Second Edition*, Elsevier, Amsterdam Netherlands, 1997, pp. 3-31.

J. R. Noseworthy, "The Test and Training Enabling Architecture (TENA)—Supporting the Decentralized Development of Distributed Applications and LVC Simulations", *Proceedings of the Twelfth IEEE/ACM International Symposium on Distributed Simulation and Real-Time Applications*, Vancouver Canada, October 27-29 2008, pp. 259-268.

H. T. Odum and E. C. Odum, *Modeling For All Scales: An Introduction to System Simulation*, Academic Press, San Diego CA, 2000.

T. I. Ören, "Rationale for a Code of Professional Ethics for Simulationists", in Society for Modeling & Simulation International, *Proceedings of the 2002 Summer Computer Simulation Conference*, San Diego CA, July 14-18 2002, pp. 428-433.

T. I. Ören, "Uses of Simulation", in J. A. Sokolowski and C. M. Banks, (Editors), *Principles of Modeling and Simulation: A Multidisciplinary Approach*, John Wiley & Sons, Hoboken NJ, 2009, pp. 153-179.

T. I. Ören, M. S. Elzas, I. Smit, and L. G. Birt, "A Code of Professional Ethics for Simulationists", in Society for Modeling & Simulation International, *Proceedings of the 2002 Summer Computer Simulation Conference*, San Diego CA, July 14-18 2002, pp. 434-435.

W. J. Palm, *Introduction to Matlab 6 for Engineers*, McGraw-Hill, Boston MA, 2001.

Y. Papelis and P. Madhavan, "Modeling Human Behavior", in J. A. Sokolowski and C. M. Banks (Editors), *Modeling and Simulation Fundamentals: Theoretical Underpinnings and Practical Domains*, John Wiley & Sons, Hoboken NJ, 2010, pp. 271-324.

E. P. Paulo, R. Jimenez, B. Rowden, and C. Causee, "Simulation Analysis of a System to Defeat Maritime Improvised Explosive Devices (MIED) in a US Port", *Journal of Defense Modeling & Simulation*, 2010, pp.115–125.

M. B. Pettitt, M. Mayo, and J. Norfleet, "Medical Simulation Training Systems", in D. Cohn, J. Nicholson, and D. Schmorrow (Editors) *The PSI Handbook of Virtual Environments for Training and Education, Volume 3: Integrating Systems, Training Evaluations, and Future Operations*, Praeger Security International, Westport CT, 2009, 99-106.

M. D. Petty, "Behavior Generation in Semi-Automated Forces", in D. Nicholson, D. Schmorrow, and J. Cohn (Editors), *The PSI Handbook of Virtual Environments for Training and Education: Developments for the Military and Beyond, Volume 2: VE Components and Training Technologies*, Praeger Security International, Westport CT, 2009, pp. 189-204

M. D. Petty, "Benefits and Consequences of Automated Learning in Computer Generated Forces Systems", *Information & Security*, Volume 12, Number 1, 2003, pp. 63-74.

M. D. Petty, "Verification and Validation", in J. A. Sokolowski and C. M Banks, (Editors), *Principles of Modeling and Simulation: A Multidisciplinary Approach*, John Wiley & Sons, Hoboken NJ, 2009, pp. 121-149.

M. D. Petty, "Verification, Validation, and Accreditation", in J. A. Sokolowski and C. M. Banks (Editors), *Modeling and Simulation Fundamentals: Theoretical Underpinnings and Practical Domains*, John Wiley & Sons, Hoboken NJ, 2010, pp. 325-372.

M. D. Petty, R. W. Franceschini, and J. Panagos, "Multi-Resolution Combat Modeling", in A. Tolk (Editor), *Engineering Principles of Combat Modeling and Distributed Simulation*, John Wiley & Sons, Hoboken NJ, 2012, pp. 607-640.

D. Popken and L. Cox, "A Simulation-Optimization Approach to Air Warfare Planning", *Journal of Defense Modeling & Simulation: Applications, Methodology, Technology*, 2004, pp. 127–140.

R. S. Pressman, *Software Engineering: A Practitioner's Approach*, McGraw-Hill, New York NY, 2001.

A. B. Pritsker, "Principles of Simulation Modeling", in J. Banks (Editor), *Handbook of Simulation: Principles, Methodology, Advances, Applications, and Practice*, John Wiley & Sons, New York NY, 1998, pp. 31–51.

T. Pulecchi, F. Casella and M. Lovera, "Object-Oriented Modelling for Spacecraft Dynamics: Tools and Applications", *Simulation Modelling Practice and Theory*, Vol. 18, Iss. 1, January 2010, pp. 63–86.

L. B. Rainey (Editor), *Space Modeling and Simulation: Roles and Applications Throughout the System Life Cycle*, The Aerospace Press, 2004.

P. F. Reynolds, A. Natrajan, and S. Srinivasan, "Consistency Maintenance in Multiresolution Simulations", *ACM Transactions on Modeling and Computer Simulation*, Vol. 7, No. 3, July 1997, pp. 386-392.

I. Ridpath, *Norton's Star Atlas 2000.0*, Longman Scientific & Technical, London, 1989, p. 132 P. F. Reynolds, "The Role of Modeling and Simulation", in J. A. Sokolowski and C. M. Banks (Editors), *Principles of Modeling and Simulation: A Multidisciplinary Approach*, John Wiley & Sons, Hoboken NJ, 2009, pp. 25-43.

M. W. Rohrer, "Simulation of Manufacturing and Material Handling Systems", in J. Banks (Editor), *Handbook of Simulation: Principles, Methodology, Advances, Applications, and Practice*, John Wiley & Sons, 1998, pp. 519-545.

K. Ross, J. Phillips, and J. Cohn, "Creating Expertise with Technology Based Training", in D. Schmorrow, J. Cohn, D. Nicholson (Editors), *The PSI Handbook of Virtual Environments for Training and Education, Volume 1: Learning, Requirements, and Metrics*, Praeger Security International, Westport CT, 2009, pp. 66-80.

M. B. Rosson and J. M. Carroll, "Expertise and Instruction in Software Development", in M. G. Helander, T. K. Landauer, and P. V. Prabhu (Editors), *Handbook of Human-Computer Interaction, Second Edition*, Elsevier, Amsterdam Netherlands, 1997, pp. 1105-1126.

R. Y. Rubinstein and D. K. Kroese, *Simulation and the Monte Carlo Method*, John Wiley & Sons, Hoboken NJ, 2008.

J. Rumbaugh, I. Jacobson, and G. Booch, *The Unified Modeling Language Reference Manual*, Addison-Wesley, Reading MA, 1999.

R. Sadek, "Audio", in D. Nicholson, D. Schmorrow, and J. Cohn (Editors), *The PSI Handbook of Virtual Environments for Training and Education, Volume 2: VE Components and Training Technologies*, Praeger Security International, Westport CT, 2009, pp. 90-115.

M. Schwartz, *Principles of Electrodynamics*, Dover, Mineola NY, 1987.

J. Scott, *Social Network Analysis: A Handbook*, Sage, London UK, 2000.

S. Sepahban, J. Aguilar, A. Dawdy, S. Paige, and J. Sercel, "Modeling and Simulation in Conceptual Design", in L. B. Rainey (Editor), *Space Modeling and Simulation: Roles and Applications Throughout the System Life Cycle*, The Aerospace Press, El Segundo CA, 2004, pp. 197-223.

M. A. Seeds, *Foundations of Astronomy*, Brooks/Cole–Thomson, Pacific Grove GA, 2003.

L. A. Segel, *Mathematics Applied to Continuum Mechanics*, Dover, 1987.

L. Setian, *Engineering Field Theory with Applications*, Cambridge University Press, Cambridge UK, 1992.

T. B. Sheridan, "Task Analysis, Task Allocation and Supervisory Control", in M. G. Helander, T. K. Landauer, and P. V. Prabhu (Editors), *Handbook of Human-Computer Interaction, Second Edition,* Elsevier, Amsterdam, Netherlands, 1997, pp. 87-105.

R. Shumaker, "Technological Prospects for a Personal Virtual Environment", in J. Cohn, D. Nicholson, and D. Schmorrow (Editors), *The PSI Handbook of Virtual Environments for Training and Education, Volume 3: Integrating Systems, Training Evaluations, and Future Operations,* Praeger Security International, Westport CT, 2009, pp. 355-370.

H. A. Simon, "Prediction and prescription in systems modeling", *Operations Research,* Vol. 38, No. 1, 1990, pp. 7-14.

R. Sinha, C. J. J. Paredis, V. C. Liang, and P. K. Khosla, "Modeling and Simulation Methods for Design of Engineering Systems", *Journal of Computing and Information Science in Engineering,* Vol. 1, 2001, pp. 84–91.

J. A. Sokolowski, "Enhanced Decision Modeling Using Multiagent System Simulation", *SIMULATION,* Vol. 79, No. 4, pp. 232-242.

J. A. Sokolowski and C. M Banks, *Modeling and Simulation for Analyzing Global Events,* John Wiley & Sons, Hoboken NJ, 2009.

J. A. Sokolowski, "Monte Carlo Simulation", in J. A. Sokolowski and C. M. Banks (Editors), *Modeling and Simulation: Theoretical Underpinnings and Practical Domains,* John Wiley & Sons, Hoboken NJ, 2010, pp. 131–145.

L. Smith, *Chaos: A Very Short Introduction,* Oxford University Press, Oxford, 2007.

R. D. Smith, *Military Simulations & Serious Games,* Modelbenders Press, Orlando FL, 2009.

I. Ståhl, "New Product Development: When Discrete Simulation is Preferable to System Dynamics", *Proceedings of the 1995 EUROSIM Conference,* Vienna Austria, September 11-15 1995, pp. 1089-1094.

J. T. Staley and G. H. Orians, "Evolution and the Biosphere" in M.C. Jacobson, R.J. Charlson, H. Rodhe, and G. H. Orians (Editors), *Earth System Science,* Elsevier, London, 2000.

J. Stewart, *Calculus, Early Transcendentals Single Variable, Fifth Edition,* Brooks/Cole, Belmont CA, 2003.

S. K. Stein, *Calculus and Analytic Geometry,* McGraw-Hill, New York NY, 1988.

M. Whitton and R. B. Loftin, "Section Perspective", in D. Nicholson, D. Schmorrow, and J. Cohn (Editors), *The PSI Handbook of Virtual Environments for Training and Education, Volume 2: VE Components and Training Technologies*, Praeger Security International, Westport CT, 2009, pp. 1-14.

Wolfram Research, *A Quick Tour of Mathematica 5*, Wolfram Media, Champaign IL, 2004. O. C. Zienkiewicz, R. L. Taylor and J. Z. Zhu, *The Finite Element Method: Its Basis and Fundamentals, Sixth Edition*, Elsevier, Amsterdam, 2005.

REFERENCES

Cannon-Bowers,J.A. and E. Salas.1998. "Making Decisions Under Stress, Implications for Individual and Team Training". Washington, DC.: American Psychological Association.

Driskell, J.E. and E. Salas.1996." Stress and Human Performance".Mahwah NJ:Lawrence Erlbaum Associates, Inc.

Hayes,R.T. and M.J. Singer.1988. "Simulation Fidelity in Training System Design". New York:Springer-Verlag.

International Atomic Energy Agency TECDOC-1411. 2004. "Use of Control Room Simulators for Training of Nuclear Power Plant Personnel".

Janosy,J.S. (2011). "Simulation and Simulators for Nuclear Power Generation, Nuclear Power -System Simulations and Operation," Dr. Pavel Tsvetkov (Ed.), ISBN: 978-953-307-506-8, InTech, Available

Kelly,L. (1970). "The Pilot Maker".New York, :Grosset and Dunlap

Myles, B.(1990). "Night Witches: The Amazing Story of Russia's Women Pilots in World War II".

NAWCTSD News. (February 1994)

Page, R.L. (Date Unknown). "Brief History of Flight Simulation".Formerly of Quantas Simulation Services: citeSeerX.

Pimental, K., and Blau, B. (1994). "Teaching Your System To Share." IEEE computer graphics and applications", 14(1), 60

Rheingold, H. (1992). "Virtual Reality", Simon & Schuster, New York, N.Y.

Rolfe, J.M.and K.J.Staples.(1998)." "Flight Simulation". Cambridge Aerospace Series. New York: Cambridge University Press.

Ulrich,W. A History of Simulation Part III—Preparing for War Halldale

Zyda,M.(Chair).1997." Modeling and Simulation, Linking Entertainment and Defense". *National Research Council.* Washington,DC: National Academy Press.

ACKNOWLEDGEMENTS

I want to thank the following individuals for their contributions to this writing:

Dick Adkins, Artist, whose imagination stimulates ours.
Vince Amico for being an inspiration to me and the simulation community.
The "Archie" Arcidiacono family for service to our country through military simulation.
Dr. Kim Dahl who reviewed the manuscript.
John Guba (USMC, Ret.), my uncle, who served his country at Guadulcanal, and taught me the value of freedom.
Dave Parker, Artist, who brings to life the past.
Mike Phillips, TRADEVMAN, who relives the history of simulator logistics.
Dr. Randy Shumaker, Research Manager, who shares a glimpse of the future with us.
Dr. Roger Smith, Futurist, who extrapolates the past and present to the future.
Walter Ulrich, Historian, of Halldale Publications who allowed the author use of the photos and material on early simulation.
Phil Wisniewski and daughter, Judy, who went over this book with a fine-toothed comb.

A special thanks to my wife, Judy, and to my children, Anna, Joseph and Andrew; my step-children Greg (Stephanie), Bryan (Wendy), Lisa-and our fourteen grandchildren.

www.ingramcontent.com/pod-product-compliance
Lightning Source LLC
LaVergne TN
LVHW070752060326
832904LV00016BA/310/J